"This book is an important contribution to the field. I have been publishing articles using EFA for over 30 years, yet it provided me with new insights and information on EFA. More importantly, the material is easy to follow and accessible to researchers and graduate students new to EFA. I highly recommend it to anyone seeking to become competent in EFA."

Joseph J. Glutting, *University of Delaware, USA*

"*A Step-by-Step Guide to Exploratory Factor Analysis with Stata* offers not only an explanation of how to use Stata but also a clear overview of how to conduct an EFA. It is a valuable resource for students and researchers alike."

Stefan C. Dombrowski, *Rider University, USA*

"This book is an amazing resource for those new to factor analysis as well as those who have used it for some time. It is a terrific guide to best practices in exploratory factor analysis with rich explanations and descriptions for why various procedures are used and equally terrific in providing resources and guidance for using Stata for conducting factor analysis. I highly recommend this book."

Gary L. Canivez, *Eastern Illinois University, USA*

"This is by far, the best book of its kind. Professor Watkins has an accessible and engaging writing style that nicely blends underlying theoretical assumptions of exploratory factor analysis with easy-to-follow, practical suggestions for the statistical calculations. This text will be a tremendous resource for beginning, intermediate, and advanced researchers. Highly recommended!"

Joseph C. Kush, *Duquesne University, USA*

A Step-By-Step Guide to Exploratory Factor Analysis with Stata

This is a concise, easy to use, step-by-step guide for applied researchers conducting exploratory factor analysis (EFA) using **Stata**.

In this book, Dr. Watkins systematically reviews each decision step in EFA with screen shots of **Stata** code and recommends evidence-based best practice procedures. This is an eminently applied, practical approach with few or no formulas and is aimed at readers with little to no mathematical background. Dr. Watkins maintains an accessible tone throughout and uses minimal jargon and formula to help facilitate grasp of the key issues users will face when applying EFA, along with how to implement, interpret, and report results. Copious scholarly references and quotations are included to support the reader in responding to editorial reviews.

This is a valuable resource for upper level undergraduate and postgraduate students, as well as for more experienced researchers undertaking multivariate or structure equation modeling courses across the behavioral, medical, and social sciences.

Marley W. Watkins earned a PhD in Educational Psychology and Measurements with a specialty in School Psychology from the University of Nebraska-Lincoln, USA. He is currently Research Professor in the Department of Educational Psychology at Baylor University, USA, and has authored more than 200 articles, books, and chapters and presented more than 150 papers at professional conferences.

A Step-By-Step Guide to Exploratory Factor Analysis with Stata

Marley W. Watkins

Routledge
Taylor & Francis Group

NEW YORK AND LONDON

First published 2022
by Routledge
605 Third Avenue, New York, NY 10158

and by Routledge
2 Park Square, Milton Park, Abingdon, Oxon OX14 4RN

Routledge is an imprint of the Taylor & Francis Group, an informa business

Library of Congress Cataloging-in-Publication Data
Names: Watkins, Marley W., 1949– author.
Title: A step-by-step guide to exploratory factor analysis
with Stata / Marley W. Watkins.
Description: New York, NY: Routledge, 2021. |
Includes bibliographical references and index.
Identifiers: LCCN 2021007759 (print) | LCCN 2021007760 (ebook) |
ISBN 9780367710996 (hardback) | ISBN 9780367710323 (paperback) |
ISBN 9781003149286 (ebook)
Subjects: LCSH: Mathematical statistics–Data processing. | Stata.
Classification: LCC QA276.4 .W38 2021 (print) |
LCC QA276.4 (ebook) | DDC 519.5/350285536–dc23
LC record available at https://lccn.loc.gov/2021007759
LC ebook record available at https://lccn.loc.gov/2021007760

ISBN: 978-0-367-71099-6 (hbk)
ISBN: 978-0-3677-1032-3 (pbk)
ISBN: 978-1-003-14928-6 (ebk)

DOI: 10.4324/9781003149286

Typeset in Bembo
by Newgen Publishing UK

Access the Support Material: www.routledge.com/9780367710323

Contents

Figures

Preface

Exploratory factor analysis (EFA) was developed more than 100 years ago (Spearman, 1904) and has been extensively applied in many scientific disciplines across the ensuing decades (Finch, 2020a). A PsychInfo database search of "exploratory factor analysis" found more than 14,000 citations for the years 2000–2020. However, EFA is a complex statistical tool that is all too easy to misapply, resulting in flawed results and potentially serious negative consequences (Preacher & MacCallum, 2003). Surveys of published research have consistently found that questionable or inappropriate EFA methods were applied (Conway & Huffcutt, 2003; Fabrigar et al., 1999: Ford et al., 1986; Gaskin & Happell, 2014; Goretzko et al., 2019; Henson & Roberts, 2006; Howard, 2016; Izquierdo et al., 2014; Lloret et al., 2017; McCroskey & Young, 1979; Norris & Lecavalier, 2010; Park et al., 2002; Plonsky & Gonulal, 2015; Roberson et al., 2014; Russell, 2002; Sakaluk & Short, 2017; Thurstone, 1937). The uniformity of these results across scientific disciplines (e.g., business, psychology, nursing, disability studies, etc.) and 80 years is striking.

Many graduate courses in multivariate statistics in business, education, and the social sciences provide relatively little coverage of EFA (Mvududu & Sink, 2013). A non-random online search for syllabi found, for example, that a graduate course in multivariate statistics at Oklahoma State University devoted three weeks to a combination of principal components analysis, EFA, and confirmatory factor analysis (CFA) whereas a multivariate course at the Graduate School of Business at Columbia University allocated two weeks to cover all forms of factor analysis. In recent decades, courses in structural equation modeling (SEM) have become popular and might include EFA as an introduction to CFA. For instance, an SEM course at the University of Nebraska devoted less than one week to coverage of EFA whereas a similar course at the University of Oregon failed to include any mention of EFA. Of course, there is no assurance that all, or most, students are exposed to even the minimal content found in these courses. A survey of colleges of education found that doctoral programs typically required only four methods courses of which the majority were basic (Leech & Goodwin, 2008). Likewise, surveys of curriculum requirements in psychology have revealed that a full course in factor analysis was offered in only 18% of doctoral psychology programs whereas around 25% offered no training at all (Aiken et al., 2008). As summarized by Henson et al. (2010), "the general level of training is inadequate for many, even basic, research purposes" (p. 238).

Researchers must make several thoughtful and evidence-based methodological decisions while conducting an EFA (Henson & Roberts, 2006). There are a number of options available for each decision, some better than others (Lloret et al., 2017). Poor decisions can produce "distorted and potentially meaningless solutions" (Ford et al., 1986, p. 307) that can negatively affect the development and refinement of theories and measurement instruments (Bandalos & Gerstner, 2016; Fabrigar & Wegener, 2012; Henson & Roberts, 2006; Izquierdo et al., 2014; Lloret et al., 2017) and thereby "create an illusion of scientific certainty and a false sense of objectivity" (Wang et al., 2013, p. 719). From a broader perspective, "understanding factor analysis is key to understanding much published research" (Finch, 2020a, p. 1) and "proficiency in quantitative methods is important in providing a necessary foundation for what many have conceptualized as scientifically based research" (Henson et al., 2010, p. 229).

In short, researchers tend to receive little formal training in EFA and as a result habitually rely on suboptimal EFA methods. Unfortunately, researchers are unlikely to make better methodological choices as they gain experience because the professional literature is littered with poor quality EFA reports that model questionable EFA methods (Plonsky & Gonulal, 2015). Additionally, researchers tend to utilize software with unsound default options for EFA (Carroll, 1978, 1983; Izquierdo et al., 2014; Lloret et al., 2017; Osborne, 2014; Widaman, 2012).

Conway and Huffcutt (2003) proposed several potential solutions to improve EFA practice. One suggestion was for "well-written books of the type that researchers are likely to turn to when conducting EFA (e.g., books on using specific software packages). These articles and books need to clearly spell out the appropriate use of EFA as well as different EFA choices and their implications and urge readers to think carefully about their decisions rather than accepting default options" (p. 166). Following that suggestion, this book systematically reviews each decision step in EFA and recommends evidence-based methodological procedures that are "markedly better than others" (Fabrigar et al., 1999, p. 294) along with the Stata commands needed to implement each recommended procedure. As such, this is an eminently applied, practical approach with few or no formulas. Rather, this book is intended to provide readers with a conceptual grasp of the main issues along with precise implementation instructions to supplement the more mathematical approach found in many multivariate and SEM books and courses. Copious scholarly references and quotations are included to provide the reader with additional resources that might be needed to respond to editorial reviews. This approach should be valuable for students as well as for more experienced researchers who wish to implement EFA in an evidence-based, best practice, scientifically defensible manner.

1 Introduction

Historical Foundations

The idea that unobservable phenomena underlie observed measurements is very old and pervasive. In fact, it may be a basic scientific principle (Hägglund, 2001). Philosophers and scientists such as Plato, Descartes, Bacon, Locke, Hume, Quetelet, Galton, Pearson, and Mill articulated these philosophical and mathematical foundations. However, it was Spearman (1904) who explicated a mathematical model of the relations between observed measures and latent or unmeasured variables (Mulaik, 1987).

Spearman (1904) described his mathematical model as a "'correlational psychology' for the purpose of positively determining all psychical tendencies, and in particular those which connect together the so-called 'mental tests' with psychical activities of greater generality and interest" (p. 205). That is, to analyze the correlations between mental tests in support of his theory of intelligence. Spearman posited a general intelligence (labeled *g*) that was responsible for the positive relationships (i.e., correlations) he found among mental tests. Given that this general intelligence could not account for the totality of the test intercorrelations, he assumed that a second factor specific to each test was also involved. "Thus was born Spearman's 'two-factor' theory which supposed that the observed value of each variable could be accounted for by something common to all variables (the general, or common, factor) and the residual (the specific factor)" (Bartholomew, 1995, p. 212). Spearman also assumed that mental test scores were measured with some degree of error that could be approximated by the correlation of two repeated measurements (i.e., test–retest reliability).

Exploratory factor analysis (EFA) methods were further debated and refined over the ensuing decades with seminal books appearing in the middle of the century (Burt, 1940; Cattell, 1952; Holzinger & Harman, 1941; Thomson, 1950; Thurstone, 1935, 1947; Vernon, 1961). These scholars found Spearman's two-factor theory over simple and proposed group factors in addition to general and specific factors. Thus, the observed value of each variable could be accounted for by something common to all measured variables (general factor), plus something common to some but not all measured variables (group factors), plus something unique to each variable (specific

DOI: 10.4324/9781003149286-1

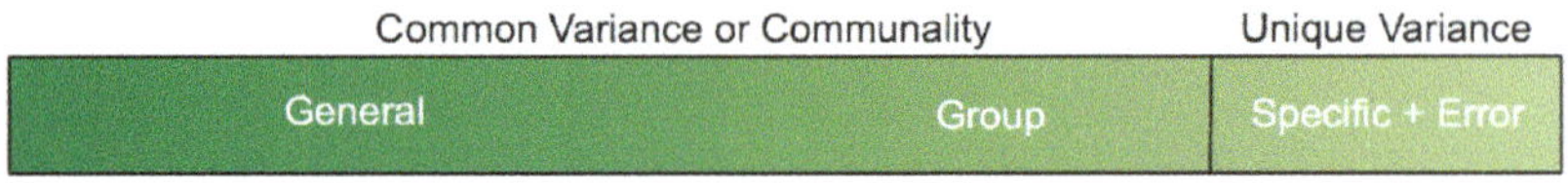

Common Variance or Communality		Unique Variance
General	Group	Specific + Error

Figure 1.1 Variance components

factor), plus error. Common variance, or the sum of variance due to both general and group factors, is called communality. The combination of specific variance and error variance is called uniqueness (Watkins, 2017). As portrayed in Figure 1.1, this is the common factor model: total variance = common variance + unique variance (Reise et al., 2018; Yong & Pearce, 2013).

The contributions of Thurstone (1931, 1935, 1940, 1947) were particularly important in the development of EFA. He studied intelligence or ability and applied factor analysis to many datasets and continued the basic assumption that "a variety of phenomena within the domain are related and that they are determined, at least in part, by a relatively small number of functional unities, or factors" (1940, p. 189). Thurstone believed that "a test score can be expressed, in first approximation, as a linear function of a number of factors" (1935, p. vii) rather than by general and specific factors. Thus, he analyzed the correlation matrix to find multiple common factors and separate them from specific factors and error. To do so, Thurstone developed factorial methods and formalized his ideas in terms of matrix algebra. Using this methodology, Thurstone identified seven intercorrelated factors that he named primary mental abilities. Eventually, he recognized that the correlations between these primary mental ability factors could also be factor analyzed and would produce a second-order general factor. Currently, a model with general, group, and specific factors that identifies a hierarchy of abilities ranging in breadth from general to broad to narrow is ascendant (Carroll, 1993).

A variety of books on factor analysis have been published. Some presented new methods or improved older methods (Cattell, 1978; Harman, 1976; Lawley & Maxwell, 1963). Others compiled the existing evidence on factor analysis and presented the results for researchers and methodologists (Child, 2006; Comrey & Lee, 1992; Fabrigar & Wegener, 2012; Finch, 2020a; Garson, 2013; Gorsuch, 1983; Kline, 1994; Mulaik, 2010; Osborne, 2014; Osborne & Banjanovic, 2016; Pett et al., 2003; Rummel, 1970; Thompson, 2004; Walkey & Welch, 2010). In addition, there has been a veritable explosion of book chapters and journal articles explicitly designed to present best practices in EFA (e.g., Bandalos, 2018; Benson & Nasser, 1998; Beaujean, 2013; Briggs & Cheek, 1986; Budaev, 2010; Carroll, 1985, 1995a; Comrey, 1988; Cudeck, 2000; DeVellis, 2017; Fabrigar et al., 1999; Ferrando & Lorenzo-Seva, 2018; Floyd & Widaman, 1995; Goldberg & Velicer, 2006; Hair et al., 2019; Hoelzle & Meyer, 2013; Lester & Bishop, 2000; Nunnally & Bernstein, 1994; Osborne et al., 2007; Preacher & MacCallum, 2003; Schmitt, 2011; Tabachnick & Fidell, 2019; Watkins, 2018; Widaman, 2012; Williams et al., 2010).

Conceptual Foundations

As previously noted, EFA is based on the concept that unobserved or latent variables underlie the variation of scores on observed or measured variables (Bollen, 2002). Alternative conceptualizations have been described by Epskamp et al. (2018). A correlation coefficient between two variables might exist due to: (a) a random relationship between those two variables, (b) one variable causing the other, or (c) some third variable being the common cause of both. Relying on the third possibility, EFA assumes that the correlations (covariance) between observed variables can be explained by a smaller number of latent variables or factors (Mulaik, 2018). "A factor is an unobservable variable that influences more than one observed measure and which accounts for the correlations among these observed measures" (Brown, 2013, p. 257).

Theoretically, variable intercorrelations should be zero after the influence of the factors has been removed. This does not happen in reality because no model is perfect and a multitude of minor influences are present in practice. Nevertheless, it is the ideal. As described by Tinsley and Tinsley (1987), EFA "is an analytic technique that permits the reduction of a large number of interrelated variables to a smaller number of latent or hidden dimensions. The goal of factor analysis is to achieve parsimony by using the smallest number of explanatory concepts to explain the maximum amount of common variance in a correlation matrix" (p. 414).

The simplest illustration of common variance is the bivariate correlation (*r*) between two continuous variables represented by the circles labeled A and B in Figure 1.2. The correlation between variables A and B represents the proportion of variance they share, which is area A•B. The amount of overlap between two variables can be computed by squaring their correlation coefficient. Thus, if $r = .50$, then the two variables share 25% of their variance.

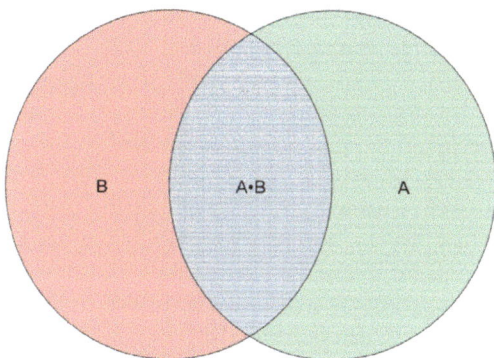

Figure 1.2 Common variance with the correlation of two variables

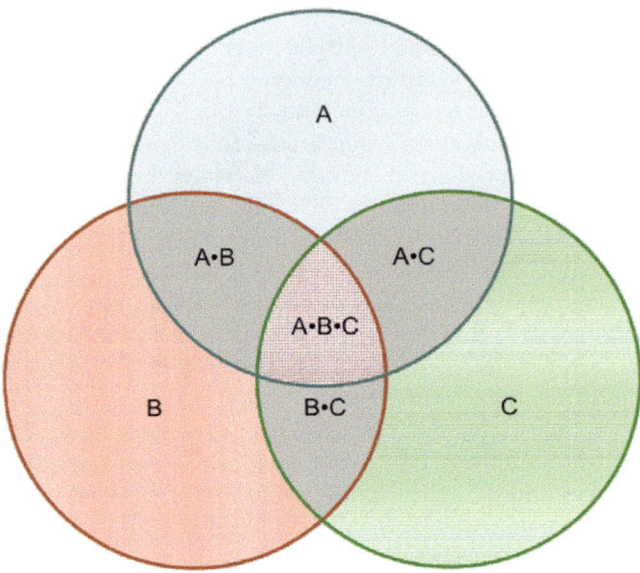

Figure 1.3 Common variance with the correlation of three variables

With three variables, the shared variance of all three is represented by area A•B•C in Figure 1.3. This represents the general factor. The proportion of error variance is not portrayed in these illustrations but it could be estimated by 1 minus the reliability coefficient of the total ABC score.

With multiple variables, the correlation structure of the data is summarized by EFA. As illustrated in Figure 1.4, there are 1 to *n* participants with scores on 1 to *x* variables (V1 to Vx) that are condensed into a V1 to Vx correlation matrix that will, in turn, be summarized by a factor matrix with 1 to *y* factors (Goldberg & Velicer, 2006).

EFA is one of a number of multivariate statistical methods. Other members of the multivariate "family" include multiple regression analysis, principal components analysis, confirmatory factor analysis, and structural equation modeling. In fact, EFA can be conceptualized as a multivariate multiple regression method where the factor serves as a predictor and the measured variables serve as criteria. EFA can be used for theory and instrument development as well as assessment of the construct validity of existing instruments (e.g., Benson, 1998; Briggs & Cheek, 1986; Carroll, 1993; Comrey, 1988; DeVellis, 2017; Haig, 2018; Messick, 1995; Peterson, 2017; Rummel, 1967; Thompson, 2004). For example, EFA was instrumental in the development of modern models of intelligence (Carroll, 1993) and personality (Cattell, 1946; Digman, 1990), and has been extensive applied for the assessment of evidence for the construct validity of numerous tests (Benson, 1998; Briggs & Cheek, 1986; Canivez et al., 2016; Watkins et al., 2002).

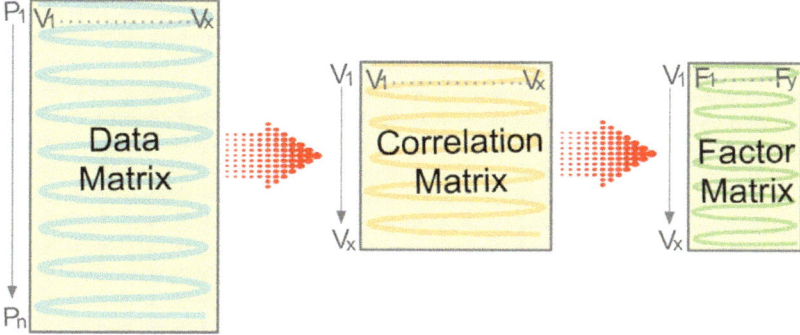

Figure 1.4 Data, correlation, and factor matrices in EFA

Readers should review the following statistical concepts to ensure that they possess the requisite knowledge for understanding EFA methods: reliability (internal consistency, alpha, test–rest, and alternate forms), true score (classical) test theory, validity (types of validity evidence), descriptive statistics (mean, mode, skew, kurtosis, standard deviation, and variance), Pearsonian correlations (product moment, phi, and point-biserial), polychoric correlation, tetrachoric correlation, partial correlation, multiple correlation, multiple regression, sample, population, multicollinearity, level of measurement (nominal, ordinal, interval, and ratio), confidence interval, and standard error of measurement. Readers can consult the textbooks written by Bandalos (2018) and Tabachnick and Fidell (2019) for exhaustive reviews of measurement and statistical concepts.

Graphical Displays and Vocabulary

Given the complexity of EFA, it is useful to display EFA models in path diagram form (Mueller & Hancock, 2019). This facilitates a quick grasp of the entire model and allows standardization of presentations. These graphs will visually illustrate the distinctions between latent variables (factors) and observed (measured) variables.

To this point, latent variables and factors have been used synonymously. Many other synonyms may be found in the professional literature, including unmeasured variables, unobserved variables, synthetic variables, constructs, true scores, hypothetical variables, and hypothetical constructs. Likewise, observed and measured variables have been used interchangeably, but terms such as manifest variables and indicator variables are also found in the professional literature. For this book, the terms factors and measured variables will be used to ensure consistency.

A simple EFA model with two factors and six measured variables is presented in Figure 1.5. In path diagrams, ellipses represent factors and

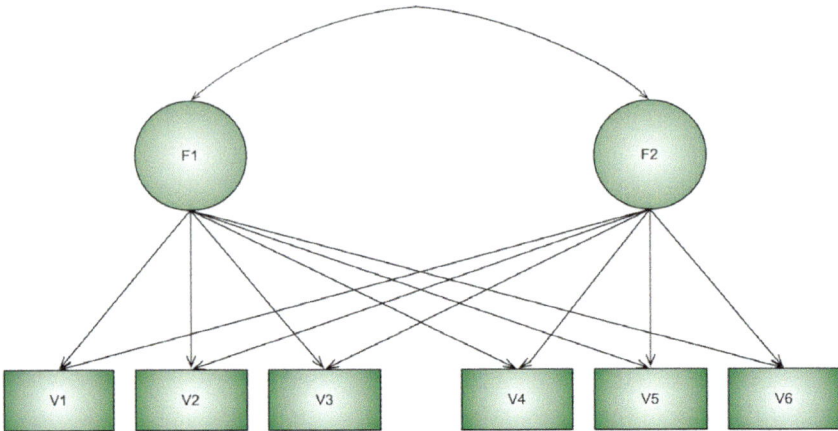

Figure 1.5 Simple EFA model with six measured variables and two correlated factors

rectangles represent measured variables. Directional relationships are indicated by single-headed arrows and non-directional (correlational) relationships by double-headed arrows. Although not included in this model, the strength of each relationship can be displayed on each directional and non–directional line. In this model, each factor directly influences all six measured variables and the two factors are correlated. Given correlated factors at a single level, this is an oblique, first-order, or correlated factors model. It represents an unrestricted solution because every factor is allowed to influence every measured variable. Path diagrams may also display error terms for measured variables, variances, etc. Although errors and variances should not be forgotten, they tend to visually clutter path diagrams and will not be displayed. Additional information about path diagrams has been provided by DeVellis (2017).

Path diagrams can also be used to conceptually illustrate the components of variance of an EFA model (Figured 1.6) with one general factor and four measured variables.

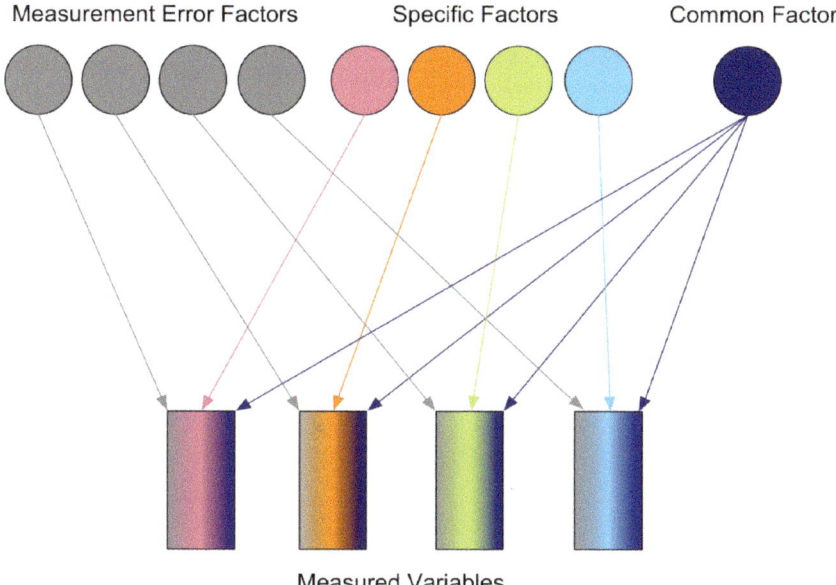

Figure 1.6 Conceptual illustration of the components of variance of an EFA model

2 Data

Six datasets are used in this book to illustrate the steps in EFA. They can be downloaded from www.routledge.com/9780367710323.

Dataset 1

The first dataset contains scores from eight tests developed to measure cognitive ability that was administered to 152 participants. This dataset is in spreadsheet format and the file is labeled iq.xlsx. These scores are from continuous variables with a mean of 100 and standard deviation of 15 in the general population.

It was anticipated that four tests would measure verbal ability (vocab1, similar1, veranal2, and vocab2) and four tests would measure nonverbal ability (designs1, matrix1, matrix2, and designs2). Brief descriptions of the eight measured variables in this iq dataset are presented in Figure 2.1 and the Pearson correlation matrix generated from the iq data is provided in Figure 2.2.

Variable Name	Description
vocab1	Provide correct definition of words
designs1	Recreate square designs with colored blocks
similar1	Describe how two words are similar
matrix1	Complete matrix designs
veranal2	Explain verbal analogies
vocab2	Recognize correct definition of words
matrix2	Recognize correct completion of matrix designs
designs2	Recreate geometric designs with puzzle pieces

Figure 2.1 Description of the variables in the iq dataset

DOI: 10.4324/9781003149286-2

	vocab1	designs1	similar1	matrix1	verana2	vocab2	matrix2	designs2
vocab1	1.00	0.58	0.79	0.62	0.69	0.82	0.56	0.51
designs1	0.58	1.00	0.57	0.65	0.51	0.54	0.59	0.66
similar1	0.79	0.57	1.00	0.60	0.70	0.74	0.58	0.55
matrix1	0.62	0.65	0.60	1.00	0.53	0.57	0.71	0.62
verana2	0.69	0.51	0.70	0.53	1.00	0.71	0.65	0.51
vocab2	0.82	0.54	0.74	0.57	0.71	1.00	0.58	0.53
matrix2	0.56	0.59	0.58	0.71	0.65	0.58	1.00	0.62
designs2	0.51	0.66	0.55	0.62	0.51	0.53	0.62	1.00

Figure 2.2 Pearson correlation matrix for the iq dataset

Dataset 2

A modification of the iq dataset was used to illustrate the procedures that can be employed with missing data. That file (iqmiss.xlsx) contains the same eight variables as Dataset 1 but missing values are indicated by values of −999.

Dataset 3

The third dataset is used to illustrate EFA with categorical variables. Each variable is one item from a self-concept scale. This dataset (sdq.xlsx) is in spreadsheet format and contains 30 variables that were administered to 425 high school students. Responses to each item ranged from 1 (*False*) to 3 (*More True Than False*) to 6 (*True*). Thus, the data are ordinal, not continuous. All negatively valanced items were reverse scored before entry and are iden-tified by an "r" appended to the item number in the dataset.

These 30 items (color-coded in Figure 2.3) are hypothesized to measure three aspects of self-concept: mathematical in red with ten items, verbal in green with ten items, and general in blue with ten items (Marsh, 1990). Reliability estimates in the .80 to .90 range have been reported for these dimensions of self-concept (Gilman et al., 1999; Marsh, 1990).

Dataset 4

The fourth dataset (HolzingerSwineford.xlsx) contains nine mental ability scales that were administered to 301 middle school students (Figure 2.4). This is an abbreviated version of the classic Holzinger and Swineford (1939) dataset that is often used to illustrate higher-order and bifactor models.

These nine scales were designed to measure three types of mental ability: spatial ability (visper, cubes, and lozenges), verbal ability (paracomp, sencomp, and wordmean), and mental speed (speedadd, speeddot, and speedcap).

No	Description	No	Description
1.	Math is my best subject	16.	Do badly on math tests
2.	Overall, I'm proud	17.	Not much to be proud of
3.	Hopeless in English class	18.	English is one of best subjects
4.	Need help in math	19.	Good grades in math
5.	Overall, I'm no good	20.	Do things as well as most
6.	Look forward to English class	21.	I hate reading
7.	Look forward to math class	22.	Never want another math course
8.	Most things I do well	23.	My life is not very useful
9.	Do badly on reading tests	24.	Good grades in English
10.	Trouble understanding math	25.	Always done well in math
11.	Nothing ever turns out right	26.	Can do almost anything if try
12.	English class is easy	27.	Trouble with writing
13.	I enjoy studying math	28.	Hate math
14.	Most things turn out well	29.	Overall I'm a failure
15.	Not good at reading	30.	Learn quickly in English class

Figure 2.3 Description of the variables in the sdq dataset

Variable	Description
visper	Visual perception
cubes	Cubes
lozenges	Lozenges
paracomp	Paragraph comprehension
sencomp	Sentence comprehension
wordmean	Word meaning
speedadd	Speeded addition
speeddot	Speeded counting of dots
speedcap	Speeded discrim of capital letters

Figure 2.4 Description of the variables in the Holzinger–Swineford dataset

No	Description	No	Description
1.	My friends think I am	11.	I have trouble with reading
2.	Read a book	12.	Reading well is
3.	Reading skill	13.	I can answer teacher questions
4.	My friends think reading is	14.	I think reading is boring-interesting
5.	Can figure out unknown	15.	For me, reading is easy-hard
6.	Tell friends about books	16.	Will spend time reading when adult
7.	Understand what I read	17.	I understand reading assignments
8.	People who read are	18.	Would like more reading time
9.	As a reader, I am	19.	When reading aloud
10.	I think libraries are	20.	Books as presents

Figure 2.5 Description of the variables in the Rmotivate practice exercise dataset

Dataset 5

The fifth dataset is the first Practice Exercise. This dataset (Rmotivate.xlsx) is also in spreadsheet format and contains 20 variables (Figure 2.5). Each variable is one item from a reading motivation scale that was administered to 500 students in grades 2 through 6 (100 at each grade level). Responses to each item ranged from 1 to 4 to represent increasingly positive opinions. For example, "My friends think I am: (1) a poor reader, (2) an OK reader, (3) a good reader, (4) a very good reader." Thus, the data are ordinal with four categories, not continuous. All negatively valanced items were reverse scored before entry.

These 20 items are hypothesized to reflect two aspects of reading motivation: reading self-concept (odd items) and value of reading (even items). Reliability estimates of .87 have been reported for each of these dimensions of reading motivation (Watkins & Browning, 2015).

An EXCEL file entitled RmotivateCorr.xlsx is also available for download. It contains Pearson coefficients in the upper diagonal and polychoric coefficients computed with Stata in the lower diagonal.

Dataset 6

The sixth dataset is used for the second Practice Exercise. This dataset (adhd.xlsx) is also in spreadsheet format and contains ten variables (Figure 2.6). Each variable is one item from a scale designed to tap the symptoms of attention–deficit hyperactivity disorder (ADHD) that was completed by 500 young adults. Respondents reported the frequency of each behavior on a four-point scale: 0 (*Never or Rarely*), 1 (*Sometimes*), 2 (*Often*), and 3 (*Very Often*). Thus, the data are ordered categories and not continuous.

Item	Description
instruct	Follow instructions
effort	Sustain mental effort
organize	Organization problems
forget	Forgetful
attention	Sustain attention
go	Constantly on the go
talks	Talk excessively
fidgets	Fidget
turns	Difficulty waiting turn
runs	Runs about

Figure 2.6 Description of the variables in the ADHD practice exercise dataset

These ten items are hypothesized to reflect two behavioral aspects of ADHD: attention problems and over-activity/impulsivity problems. It was assumed that the first five items would tap the attention problems dimension, whereas the final five items would tap the over-activity/impulsivity dimension. Similar scales with 15 to 20 items have typically found internal consistency reliability coefficients of around .85 to .90 for these factors (Nichols et al., 2017).

Figure 2.7 contains Pearson coefficients in the upper diagonal and polychoric coefficients computed with Stata in the lower diagonal for the ADHD data.

	instruct	effort	organize	forget	attention	go	talks	fidgets	turns	runs
instruct	1.00	.60	.61	.49	.57	.35	.35	.33	.42	.40
effort	.71	1.00	.56	.50	.59	.37	.33	.33	.40	.38
organize	.71	.64	1.00	.53	.54	.36	.38	.31	.46	.40
forget	.58	.58	.61	1.00	.47	.40	.33	.38	.38	.36
attention	.70	.68	.63	.53	1.00	.51	.40	.49	.48	.50
go	.42	.43	.41	.45	.60	1.00	.51	.67	.49	.54
talks	.44	.38	.43	.37	.48	.70	1.00	.43	.50	.44
fidgets	.43	.38	.37	.44	.58	.74	.52	1.00	.43	.48
turns	.50	.49	.54	.44	.56	.61	.63	.55	1.00	.45
runs	.51	.46	.50	.44	.61	.68	.58	.60	.59	1.00

Figure 2.7 Pearson (upper) and polychoric (lower) correlation matrices for the ADHD practice exercise dataset

Correlations have been reported to two decimal places because statistical simulations have shown that "it never makes sense, unless one has a sample size greater than 100,000, to report results beyond the first two leading digits" (Bedeian et al., 2009, p. 693).

3 Stata Software

There are many software packages that can be used to conduct exploratory factor analysis (EFA). SPSS, SAS, and Stata are the most popular commercial packages. A review of statistical software used in academic articles found that SPSS was cited most frequently, followed by **R**, SAS, and Stata (see http://r4stats.com/articles/popularity). Stata is frequently licensed by colleges and universities so it is widely available to faculty and students. Stata is also accessible via group license and for single users in business, government, and education. Accordingly, Stata will be employed in this book.

Stata

StataCorp released the first version of Stata in 1985 and the seventeenth version in 2021. Stata is available for Windows, Macintosh, and Linux operating systems with versions that differ by number of computer cores, variables, and participants they can manage but not by the statistical methods they include. This book will reflect the operation of Stata 16 on a Macintosh computer and will use the generic title of Stata regardless of version or operating system. There may be minor differences between operating systems and versions of Stata but results will generally be similar. Detailed information about the versions of Stata are available at stata.com.

Users should download and install Stata according to instructions from StataCorp and their sponsoring college, university, or employer. Once installed, Stata can be launched like other programs.

Stata is a command-driven program that also provides a menu system. The commands in Stata are based on an underlying programming language that can be used to perform matrix calculations and to add new features to Stata. Knowledgeable users can create new commands and, if desired, make them available to other users through online communities. However, those contributed commands cannot be accessed through the Stata menu system.

Commands may be easier than the Stata graphical user interface, at least in the long run. First, command code is relatively easy to type (or paste from a separate document or from an analysis window), whereas the point-and-click method may require navigation through multiple menus. Second, command code can easily be edited and rerun if an error is found or new participants

DOI: 10.4324/9781003149286-3

added to the data, whereas the point-and-click method will require repeated navigation through all the menus. Third, the commands can be saved in case there is a future question about the analyses. This saved record can facilitate scientific replication. Finally, some Stata options may not be available in the menu system and must be implemented via commands. That being said, it may be more convenient to use menus for data input and manipulation and easier to use commands when conducting analyses. Both methods will be illustrated in this book to ensure that users have multiple options.

Stata Windows

After launching, a **Stata** window that contains a menu bar, shortcut icons, and resizable panes is opened as displayed in Figure 3.1. Stata panes and windows are denoted by **bold** type and menu options are designated by ***bold italic*** type in this book. The **Stata** menu bar contains ***File***, ***Edit***, ***View***, ***Data***, ***Graphics***, ***Statistics***, ***User***, ***Window***, and ***Help*** menus.

The working directory is identified in the status bar below the **Command** pane. Stata looks for data within the working directory so it is a good idea to set the working directory for each project so that those files are located in the same directory (***File > Change working directory***). Stata can create a

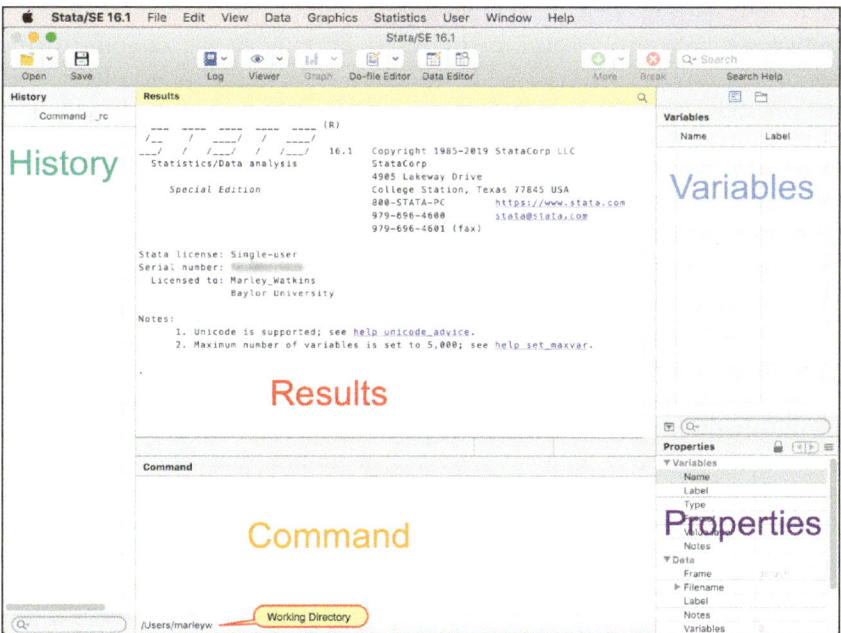

Figure 3.1 Stata window

project (*File > New > Project*) that will automatically set the working directory to where the project manager is saved.

The default **Stata** window contains multiple resizable and moveable panes. The **Command** pane is where Stata commands are entered. Previously executed commands can be sequentially recalled by pressing the PAGE UP key. Upon execution, commands entered in the **Command** pane and those selected via a menu are automatically displayed in the **Results** pane along with the results of those commands. Commands will be displayed in the Letter Gothic Std font in this book.

The **Results** pane sequentially displays Stata commands and their results. Contents of the **Results** pane can be selected and copied via *Edit > Copy* or *Edit > Copy Table* and pasted to a word processor document. In that case, the text may be misaligned in the word processor document, necessitating a change of font type and size for proper alignment (e.g., Courier 9-point). Additionally, a picture of the selected results can be obtained via *File > Copy as picture* menu options.

Alternatively, a text file can be created from the **Results** pane by creating a log file (*File > Log > Begin*), implementing the desired commands, terminating the log function (*File > Log > Suspend*), and then opening that log file with a word processor. If desired, the *Log* short-cut icon can be used instead of the menu system. This system will preserve a record of the analysis session that might be useful if questions arise in the future or if replication is required.

The **Variables** pane contains a list of the variables in the active dataset. Each variable from that list can be inserted into the **Command** pane by double-clicking on its name or by single-clicking in the column to the left of the variable name. The **Properties** pane contains detailed information about the dataset (file name, number of variables, and number of participants) and its variables (variable name, type, format, and value label). These properties are locked by default, but can be modified by clicking the lock icon.

Variable properties can also be edited via the **Stata** *> Window > Variables Manager* menu options. As seen in Figure 3.2, the **Variables Manager** window allows variable names, labels, type, format, and value labels to be viewed and modified. Additionally, a variable can be deleted from the datafile by selecting it and pressing the DELETE key.

Stata allows each variable to have both a name and a label. Variable names are usually short whereas variable labels can be longer and more descriptive. Stata recognizes three different types of integer numeric data (byte, int, and long), two types of non-integer data (float and double), and one generic type of string data (strL). Values for byte variables can range from -127 to 100, values for int variables can range from $-32,767$ to $32,740$, and values for long variables are essentially unlimited. Likewise, values for float and double variables can be quite large. Floats have about seven digits of accuracy, whereas doubles have around 16 digits of accuracy. Stata will make all calculations in double or quad precision regardless of the type of data. Strings of the strL

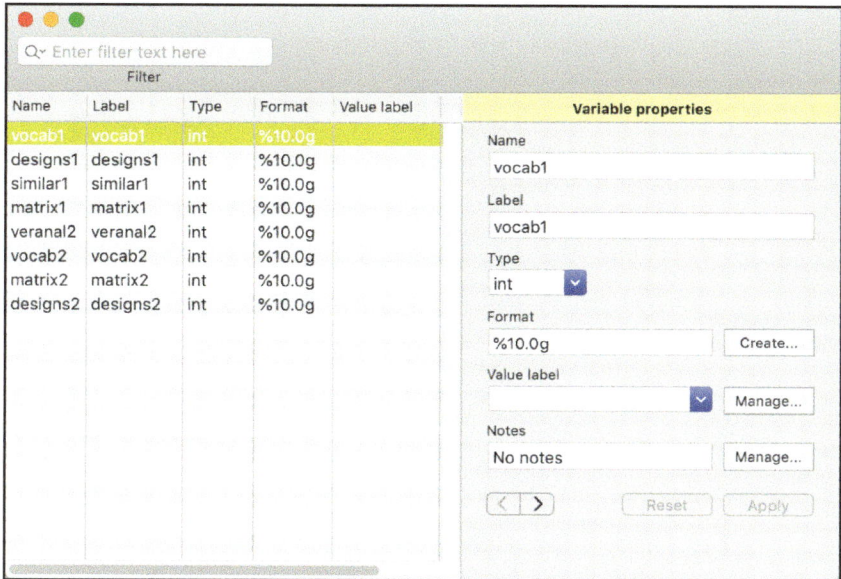

Figure 3.2 Stata variables manager window

type can be millions of characters long, but the maximum length of string variables can be mandated by str8, str16, etc. where the number identifies the maximum string length. Given the potentially large size of some variables, Stata can optimize the data with the ***Data > Data utilities > Optimize variable storage*** menu options or the compress command. It is important that the variable type be consistent with the values contained within a variable or unexpected outcomes might result.

If not included in the default display, it will be useful to add a fifth pane via ***View > Layout > Widescreen*** menu options. This new **History** pane contains a running history of all the Stata commands that have been implemented during the current session. Each of the commands in the **History** pane is a shortcut to the **Command** pane: clicking on the command in the History pane will automatically insert that command into the **Command** pane.

Many aspects of the Stata display can be modified by "Preferences" that are located in the usual Windows or Macintosh location (***Stata > Preferences***). The default preferences are probably adequate for most users, but standard fonts may be more functional than the default fonts applied by Stata. For example, using a monospace font like Courier and color-coding results and commands in the **Results** pane as displayed in Figure 3.3. Thus, the **Results** pane will display output in black 12–point Courier font and commands in blue 12–point Courier font. However, some Stata system settings are only available through commands (see Figure 3.6 for several examples).

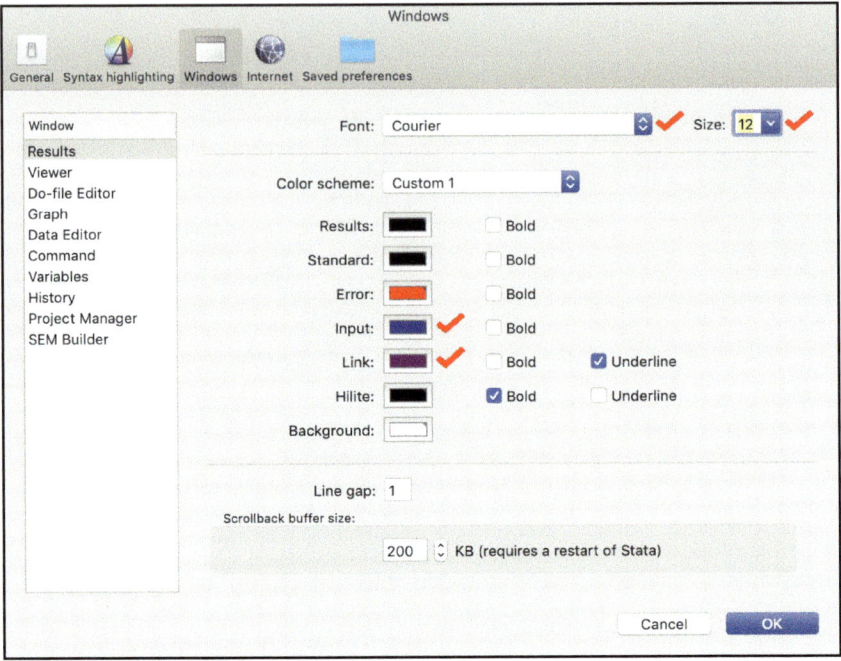

Figure 3.3 Stata preferences window

Four additional windows are available for specialized purposes. First, a **Graph** window to display graphical output. Second, a **Viewer** window to display information and tutorials. Third, the **Data Editor** window, which is similar to an Excel spreadsheet, where rows and columns of data can be entered. Traditionally, rows are individual participants and columns are variables. The **Data Editor** window can be used to view data if its *Browse* function is selected or to input and edit data if its *Edit* function is selected. Generally, data is input via the keyboard or by loading or importing from a file. Finally, the **Do-file Editor** window to create and execute "do-files" that contain a command or set of commands. Do-files can include all the commands executed during a work session and thereby serve as a record of those analyses, the basis for new analyses, or replication of those analyses. These windows can be opened by clicking a shortcut icon on the **Stata** window (Figure 3.1) or by menu options within the **Stata >** *Window* menu.

Stata Commands

Similar to the syntax or script in other software packages, commands instruct Stata to perform tasks. Stata commands are lower-case and have a typical structure of `command name variable(s) if, options`. Frequently used commands include `list, summarize, tabulate, factor, graph`, etc. A command name must be followed by one or more variable names from the active dataset, separated by spaces. For example, `summarize vocab1 designs1 similar1 matrix1`. The variable list can be shortened in the case of adjacent variables: `summarize vocab1-matrix1`. The if statement allows selection of variables based on their contents (`summarize vocab1 if vocab2 == 100`). Note that equality commands in Stata are denoted with ==, not the single = sign. Finally, each command has a set of options that control what is done and how the results are displayed. Syntax for arithmetic, logical, and relational commands are detailed in Figure 3.4.

The `display` command within the **Command** pane can be used as a calculator (`display 3^2`) or to display the value of a specific datapoint (`display vocab1[1]`). More complex operations can be computed with a combination of `generate` and `display` commands. For example, the difference between the value of the first participant's score on the vocab1 variable and the second participant's score on that same variable would be computed with `generate diff = vocab1[1] - vocab1[2]` and displayed in the **Results** pane by `display diff`. Note that the = sign was used in this mathematical computation and that the new diff variable is listed in the **Variables** pane.

Even more complex functions can be created with the `egen` command, which can operate on groups of observations and variable lists. For example, the minimum value of each observation within a row of variables could be saved as: `egen low = rowmin(vocab1-designs2)`. This command creates a new variable named low that is added to the **Variables** pane. The value of that new variable for the first participant can be shown in the

```
                                                  Relational
         Arithmetic              Logical      (numeric and string)
      ─────────────────      ─────────────    ──────────────────────
      +   addition           &    and         >    greater than
      −   subtraction        |    or          <    less than
      *   multiplication     !    not         >=   > or equal
      /   division           ~    not         <=   < or equal
      ^   power                               ==   equal
      −   negation                            !=   not equal
      +   string concatenation                ~=   not equal

      A double equal sign (==) is used for equality testing.

      The order of evaluation (from first to last) of all operators is ! (or ~),
      ^, − (negation), /, *, − (subtraction), +, != (or ~=), >, <, <=, >=, ==,
      &, and |.
```

Figure 3.4 Stata syntax for arithmetic, logical, and relational commands

Results pane by the `display low` command. All the values of that new variable can be shown with the `list low` command. Type `help egen` to see the wide variety of operators that can be employed with the `egen` command.

Options vary from command to command and can be perused via the **Help** menu or a `help` command. For instance, `help summarize` will open a new **Viewer** window with a description of the command and its options (Figure 3.5). Note that commands can be abbreviated (the minimum abbreviation is underlined in the help display). For example, `summarize`

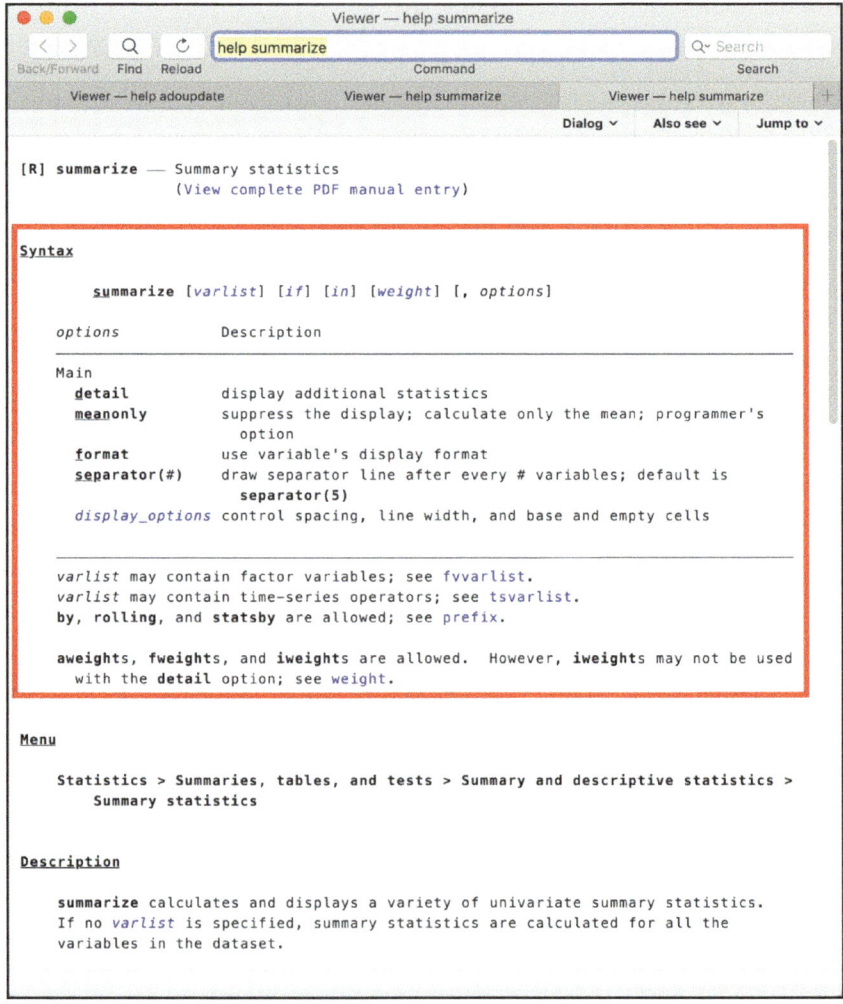

Figure 3.5 Stata help window

can be abbreviated to su. The summarize command can display additional statistics with the detail option. Thus, the full command would be: summarize vocab1 vocab2, detail. This will generate additional descriptive statistics (i.e., skew, kurtosis, and percentiles in addition to mean, standard deviation, minimum, and maximum).

Several commands that might be useful are displayed in Figure 3.6. The set commands control various system preferences that are not available through the *Preferences* menu. The cls, update all, and adoupdate commands are particularly valuable.

Commands versus Do-files

Single commands are entered in the **Command** pane of the **Stata** window and are executed when the ENTER key is pressed. In contrast, multiple commands can be sequentially executed from a do-file. The ability to execute multiple commands and maintain a record of analyses for later use or replication is a powerful feature. Using the **Do-file Editor** in conjunction

Command	Purpose
query	Review system settings
set linesize #	Number of characters per line
set pagesize #	Number of lines per page
set level #	Confidence interval level (e.g., 99)
set more off	Results pane to display results without pagination
set more on	Results pane to display results with pagination
,permanently	Set command suffix to make that setting permanent
cls	Clear the **Results** pane
clear	Erase all data from working memory
drop *Varlist*	Erase one or more variables from working memory
clear results	Erase all results from working memory
pwd	Identify the working directory
dir	Display file names in the working directory
describe	Describes the dataset currently in memory
codebook	Displays information on all the variables
update all	Update the Stata program
adoupdate, update	Update the installed ado-programs
adopath	Location of ado files

Figure 3.6 Useful stata commands

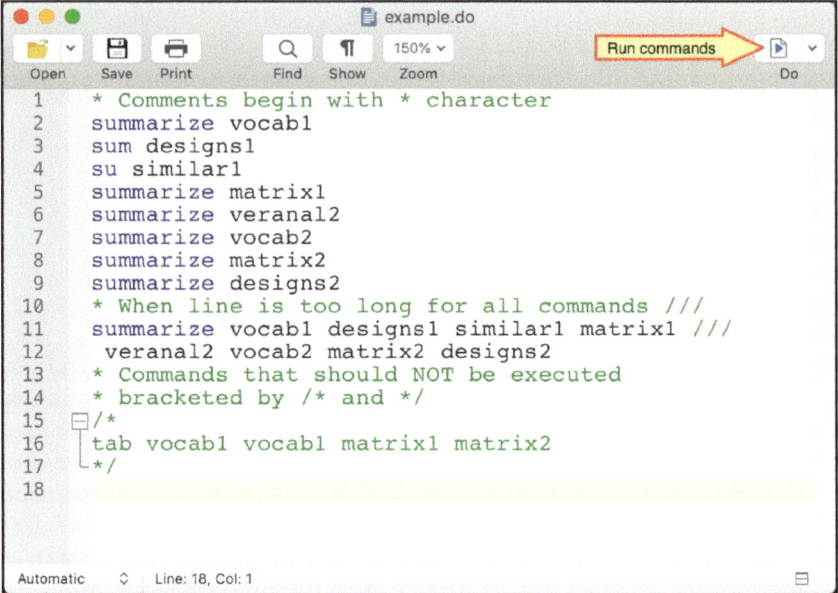

Figure 3.7 Stata do-file editor

with the **Command** pane can be an efficient system. For example, each command can be executed from the **Command** window to ensure that it operates as expected and then those valid commands can be copied from the **Command**, **Results**, or **History** pane and pasted into a **Do-file Editor** document for sequential execution. This process allows an entire analysis to be recorded in a single document.

As illustrated in Figure 3.7, the format of do-files includes the ability to enter non-executable comments and commands. All the commands in the do-file will be executed when the shortcut ***Run commands*** icon is clicked. Alternatively, selecting one or more commands by clicking on them in the **Do-file Editor** will restrict command execution to just those commands that are highlighted.

Stata Files

Stata has five native file types: data, do, graph, ado, and log. The .dta extension identifies a Stata data file. Data files can be opened and saved from the *File* menu and the shortcut icons. In addition to opening Stata data files with the *File > Open* options, Stata can import other data formats via the *File > Import* options. Specifically, Excel, text, SPSS, SAS, etc. files can be imported. Do-files are identified by the .do extension, graph files by the .gph extension, ado files by the .ado extension, and log files by either .smcl or .log extensions for formatted or plain text, respectively. As with data, the contents of the **Do-file Editor** window can be saved via the ***File > Save As*** menu or the ***Save*** icon shortcut. Stata automatically recognizes that this is a do-file and appends .do to the file name.

Ado-files are text files that contain Stata commands that have been created by users and contributed to online communities for general use. Similar to syntax files in SPSS or packages in R, these files expand Stata's capabilities. Ado-files must be downloaded and installed before they can be employed by Stata. A downloadable ado-file repository is maintained by the Statistical Software Components Archive (SSC) at the Boston College Department of Economics that can be searched with Stata commands. For example, `ssc hot, n(25)` lists the 25 most popular ado-files and `ssc whatsnew` lists the most recent ado-files at SSC. Both commands allow the listed ado-files to be downloaded and installed via online links. If an ado-file name is known, it can be directly downloaded and installed. For example, the omega ado-file can be installed with the `ssc install omega` command and it can be uninstalled with the `ssc uninstall omega` command. General information about ado-files is available via the `help ssc` command.

Ado-file names can be located through Stata's help system or by an online search implemented with the Stata `net search keyword` command where the keyword can be general (`efa`) or specific (`omega`). For example, a search for omega ado-files found three packages and two references (Figure 3.8).

Selecting the link of the ado-file that calculates a reliability estimate will generate a **Viewer** window that allows that ado-file to be installed (Figure 3.9). Detailed information about the omega ado-file can be obtained with the `help omega` command.

The computer location of Stata files (base application as well as ado-files) can be ascertained with the `sysdir` command. A list of installed ado-files can obtained with the `ado dir` command and their location can be displayed with the `adodpath` command. The `adoupdate` command lists all installed ado-files and identifies those that need to be updated, and the `adoupdate, update` command updates all installed ado-files. The source code of an ado-file can be examined via `viewsource name.ado`.

Help

Information about the many procedures available through Stata can be obtained via the ***Help*** menu, including more than 15,000 pages in the *Stata Reference Manual*, video tutorials, etc. Stata provides a general guide to obtaining help with the `help advice` command. Details about specific statistical topics can be obtained with a `help` command, followed by the name of procedure. For example, `help factor` will open a **Viewer** window with extensive details on factor analysis. Additionally, a help icon **[?]**

```
Results
. net search omega
(contacting http://www.stata.com)

3 packages found (Stata Journal and STB listed first)
-------------------------------------------------------

st0499 from http://www.stata-journal.com/software/sj17-4
    SJ17-4 st0499. Command for estimation of treatment... / Command for
    estimation of treatment effects in / the presence of correlated
    neighborhood / interactions / by Giovanni Cerulli, CNR-IRCrES, National /
    Research Council of Italy, Research / Institute on Sustainable Economic

omega from http://fmwww.bc.edu/RePEc/bocode/o  ◄——select this link to download
    'OMEGA': module to calculate the omega reliability coefficient / -omega-
    calculates McDonald's omega to estimate scale / reliability. Omega has
    many desirable statistical properties that / make it preferable to the
    widely used Cronbach's alpha.  / KW: omega / KW: alpha / KW: reliability

omega2 from https://stats.idre.ucla.edu/stat/stata/ado/analysis
    omega2.  Program to compute omega squared after an ANOVA / Philip B. Ender
    / Statistical Computing and Consulting / UCLA Office of Academic Computing
    / ender@ucla.edu / STATA ado and hlp files in the package /
    distribution-date: 20071120


2 references found in tables of contents
-----------------------------------------

http://fmwww.bc.edu/RePEc/bocode/o/
    module to compute the Blinder-Oaxaca decomposition / module to compute
    decompositions of outcome differentials / module to compute the
    Blinder-Oaxaca decomposition / module to identify differences in values
    across observations for a variable / module to display observations of

https://stats.idre.ucla.edu/stat/stata/ado/analysis/
    Welcome to UCLA Academic Technology Services Stata programs. / These
    programs include tools for data analysis. / These include programs from
    the Stata Technical Bulletin, / courtesy of, and copyright, Stata
    Corporation. / For more information about these programs, see / our web
```

Figure 3.8 Search for Stata ado-file

is included on many analysis menu windows. A general search for a given topic can be implemented with the `search` command followed by logical keywords. For example, `search factor analysis` will open a Viewer window with links to factor analysis topics.

Introductory texts that cover basic Stata operation might be useful for inexperienced users (Acock, 2018; Longest, 2019). There are also books for experienced statistical users (Baldwin, 2019; Mehmetoglu & Jakobsen, 2017). Texts that focus on Stata data management (Mitchell, 2020) and programming are also available (Baum, 2016). A wide variety of more advanced texts are published by Stata Press.

There is considerable online support for Stata users. First, Stata provides extensive support, including video tutorials, manuals, blog posts, etc. at www.stata.com/support. An online Statalist Forum includes discussions about all

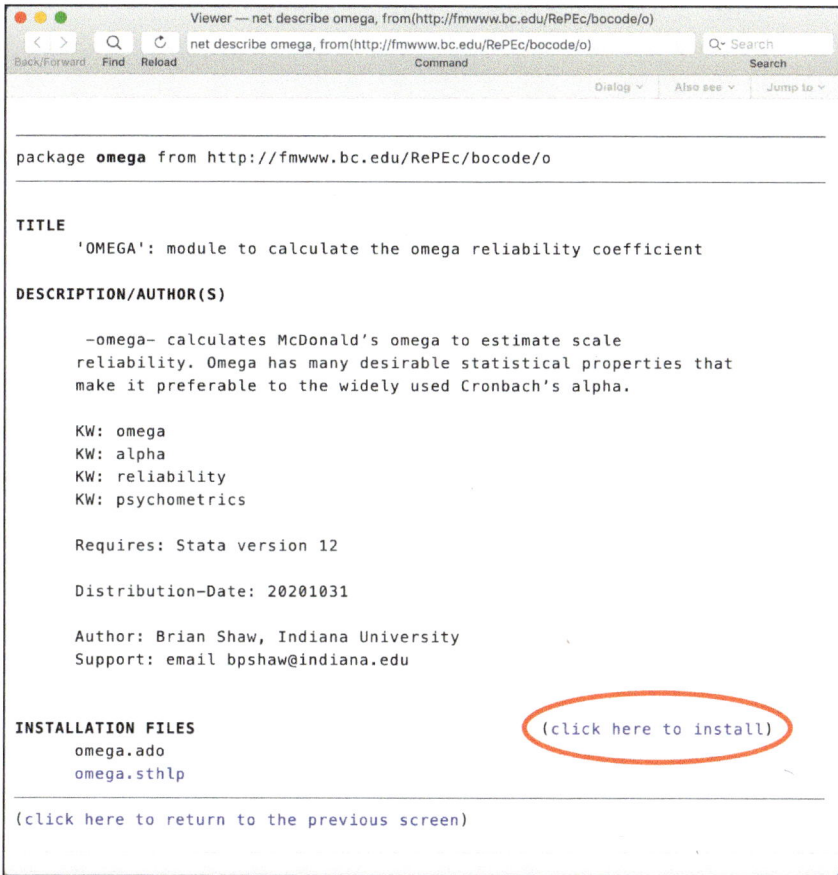

Figure 3.9 Install a Stata ado-file

aspects of Stata as well as the opportunity to ask questions of Stata users. Other online sources for information about Stata include:

www.techtips.surveydesign.com.au/blog/categories/stata
https://stats.idre.ucla.edu/stata
http://dawnteele.weebly.com/uploads/2/4/9/3/24931233/the_stata_
   bible_2.pdf
https://sociology.fas.harvard.edu/need–help–basic–stata
http://wlm.userweb.mwn.de/Stata
www.reed.edu/psychology/stata/index.html
http://homepages.rpi.edu/~simonk/pdf/UsefulStataCommands.pdf
https://libguides.library.nd.edu/data–analysis–stata
http://geocenter.github.io/StataTraining/portfolio/01_resource/
https://sites.google.com/site/mkudamatsu/stata

https://sites.tufts.edu/datalab/learning-statistics/stats-online-tutorials/
    stata-resources/
https://haghish.com/home.php
https://ideas.repec.org/s/boc/bocode.html

Registered users can email questions to Stata's technical support at tech-support@stata.com. Information about that service is available at www.stata.com/support/tech-support/contact.

4 Importing and Saving Data

Importing Data

Stata can import data via its *File > Import* menu, including Excel, text, SPSS, SAS, etc. files. For the iq data in spreadsheet format, select *File > Import > Excel* and *Browse* to find the Excel file (Figure 4.1).

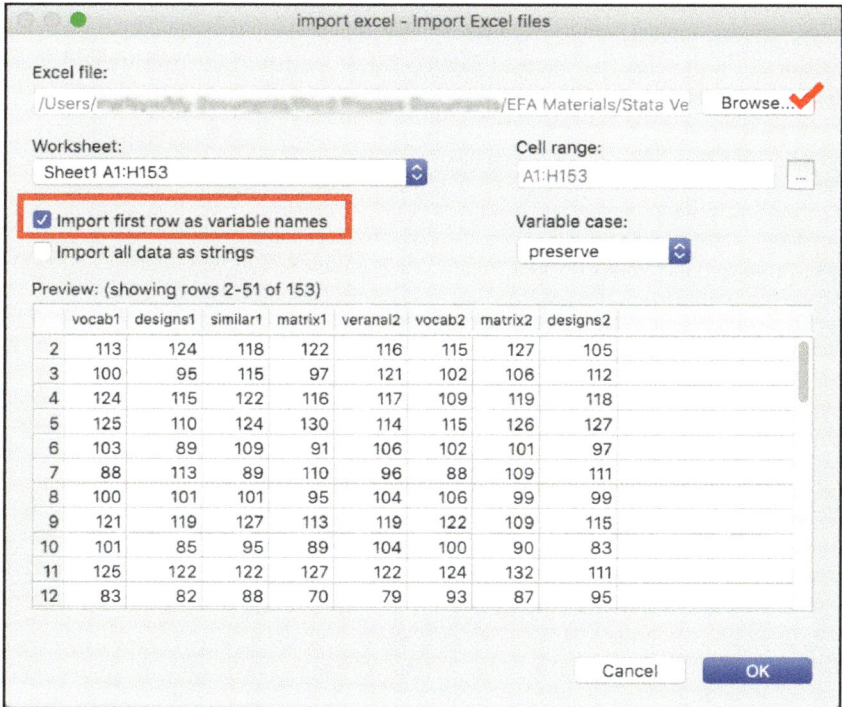

Figure 4.1 Import Excel data into Stata

DOI: 10.4324/9781003149286-4

Once the Excel file has been selected, Stata will recognize the spreadsheet format, including the possibility that variable names occupy the first row of the file. Variable names should start with a letter or underscore and contain no embedded blanks. A preview of the data is presented to ensure that the correct data and format have been selected. When satisfied, click the ***OK*** button on the bottom right corner of the import screen. Stata immediately reads that data into its memory and reports that action in the **Results** pane of the **Stata** window.

To ensure that the correct data have been imported, open the **Data Editor** window via the **Stata >** ***Window*** **>** ***Data Editor*** or the **Stata >** ***View*** **>** ***Data Editor*** **>** ***Edit*** menu options, click the ***Data Editor*** shortcut icon on the **Stata** window, or type the `Edit` command (`Browse` displays the data, `Edit` allows the data to be edited). Once open (Figure 4.2), scroll through the iq data to verify that the number of participants and number of variables correspond to the known dimensions of the iq datafile (i.e., 152 participants and 8 variables). The properties of the variables can be checked through the **Stata >** ***Window*** **>** ***Variables Manager*** menus. Details about the entire set of variables can be obtained with the `describe` and `codebook` commands.

Alternatively, single variables can be examined by adding the variable name to the `codebook` command. For example, `codebook vocab1` produces the results displayed in Figure 4.3. It is important that variables are

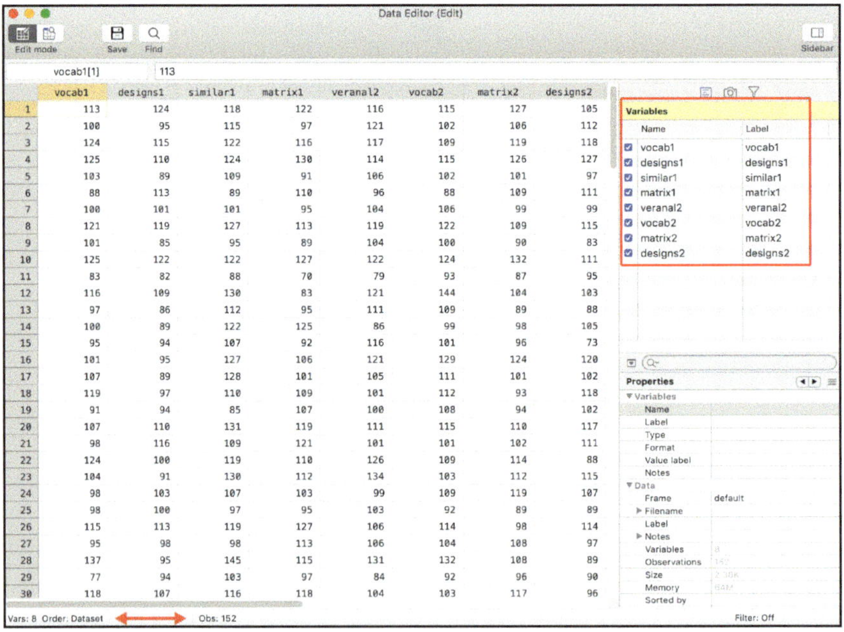

	vocab1	designs1	similar1	matrix1	veranal2	vocab2	matrix2	designs2
1	113	124	118	122	116	115	127	105
2	100	95	115	97	121	102	106	112
3	124	115	122	116	117	109	119	118
4	125	110	124	130	114	115	126	127
5	103	89	109	91	106	102	101	97
6	88	113	89	110	96	88	109	111
7	100	101	101	95	104	106	99	99
8	121	119	127	113	119	122	109	115
9	101	85	95	89	104	100	90	83
10	125	122	122	127	122	124	132	111
11	83	82	88	70	79	93	87	95
12	116	109	130	83	121	144	104	103
13	97	86	112	95	111	109	89	88
14	100	89	122	125	86	99	98	105
15	95	94	107	92	116	101	96	73
16	101	95	127	106	121	129	124	120
17	107	89	128	101	105	111	101	102
18	119	97	110	109	101	112	93	118
19	91	94	85	107	100	108	94	102
20	107	110	131	119	111	115	110	117
21	98	116	109	121	101	101	102	111
22	124	100	119	110	126	109	114	88
23	104	91	130	112	134	103	112	115
24	98	103	107	103	99	109	119	107
25	98	100	97	95	103	92	89	89
26	115	113	119	127	106	114	98	114
27	95	98	98	113	106	104	108	97
28	137	95	145	115	131	132	108	89
29	77	94	103	97	84	92	96	90
30	118	107	116	118	104	103	117	96

Figure 4.2 Stata data editor window

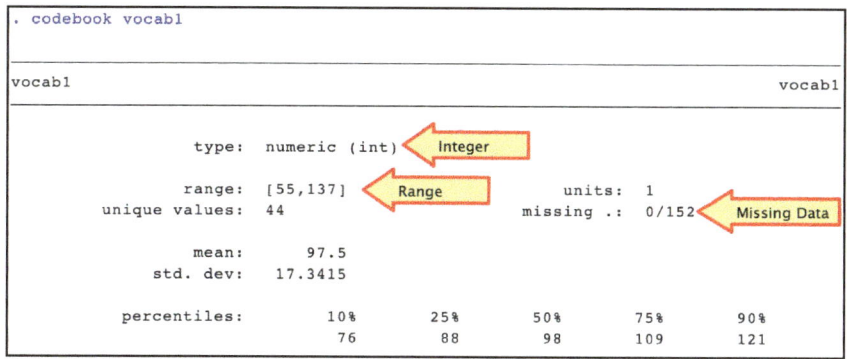

Figure 4.3 Codebook display for one variable

categorized correctly as to type (byte, int, and long integer numeric; float and double numeric data; and string data) because assumptions about measurement level may be foundational for some analyses. In this case, the vocab1 variable has been categorized by Stata as a numeric variable of integer type with values of 55 to 137 without any missing data.

Importing a Correlation Matrix

Raw data are not always available. For example, journal articles may provide the correlation matrix but not the raw data as in Figure 4.4. Correlation matrices can be input for later use in EFA with command code via a do-file window via *File > New > Do-file* menus (Figure 4.5).

Correlation matrices can also be entered using Stata's menu system. The *Statistics > Multivariate analysis > Factor and principal component analysis > Factor analysis of a correlation matrix* options will open a **factormat – Factor analysis of a matrix** window (Figure 4.6).

	vocab1	designs1	similar1	matrix1	verana2	vocab2	matrix2	designs2
vocab1	1.00	0.58	0.79	0.62	0.69	0.82	0.56	0.51
designs1	0.58	1.00	0.57	0.65	0.51	0.54	0.59	0.66
similar1	0.79	0.57	1.00	0.60	0.70	0.74	0.58	0.55
matrix1	0.62	0.65	0.60	1.00	0.53	0.57	0.71	0.62
verana2	0.69	0.51	0.70	0.53	1.00	0.71	0.65	0.51
vocab2	0.82	0.54	0.74	0.57	0.71	1.00	0.58	0.53
matrix2	0.56	0.59	0.58	0.71	0.65	0.58	1.00	0.62
designs2	0.51	0.66	0.55	0.62	0.51	0.53	0.62	1.00

Figure 4.4 Pearson correlation matrix of iq variables

```
 ●  ●  ●                              📄 InputCorrMat.do
   📂  ⌄    💾    🖨        Q    ¶    142% ⌄                                                    ▶  ⌄
  Open    Save  Print        Find  Show   Zoom                                                  Do
   1    * Input correlation matrix
   2    * Use /// to indicate that line continues
   3    matrix C = ( 1.00,0.58,0.79,0.62,0.69,0.82,0.56,0.51 \ ///
   4    0.58,1.00,0.57,0.65,0.51,0.54,0.59,0.66 \ ///
   5    0.79,0.57,1.00,0.60,0.70,0.74,0.58,0.55 \ ///
   6    0.62,0.65,0.60,1.00,0.53,0.57,0.71,0.62 \ ///
   7    0.69,0.51,0.70,0.53,1.00,0.71,0.65,0.51 \ ///
   8    0.82,0.54,0.74,0.57,0.71,1.00,0.58,0.53 \ ///
   9    0.56,0.59,0.58,0.71,0.65,0.58,1.00,0.62 \ ///
  10    0.51,0.66,0.55,0.62,0.51,0.53,0.62,1.00 )
  11    * EFA commands follow
  Automatic   ⇕   Line: 12, Col: 1
```

Figure 4.5 Import correlation matrix via the do-file editor

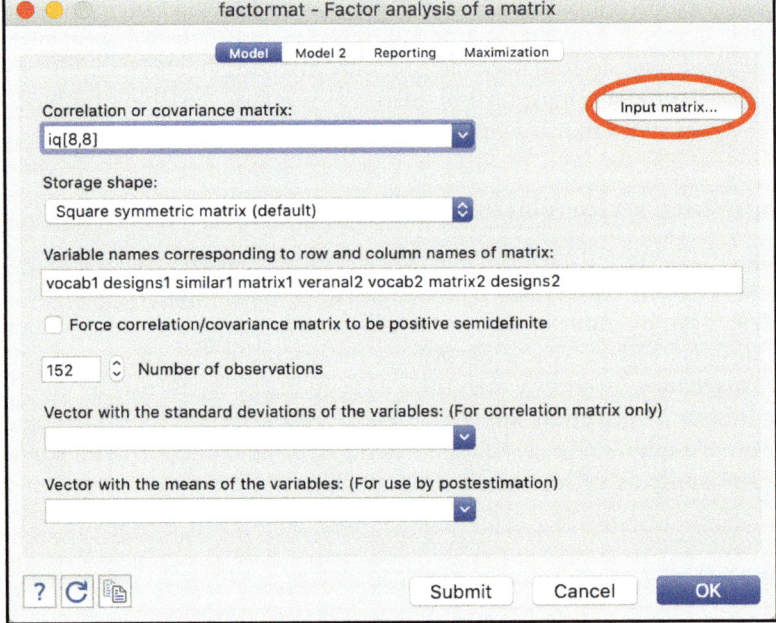

Figure 4.6 Stata factor analysis of a matrix window

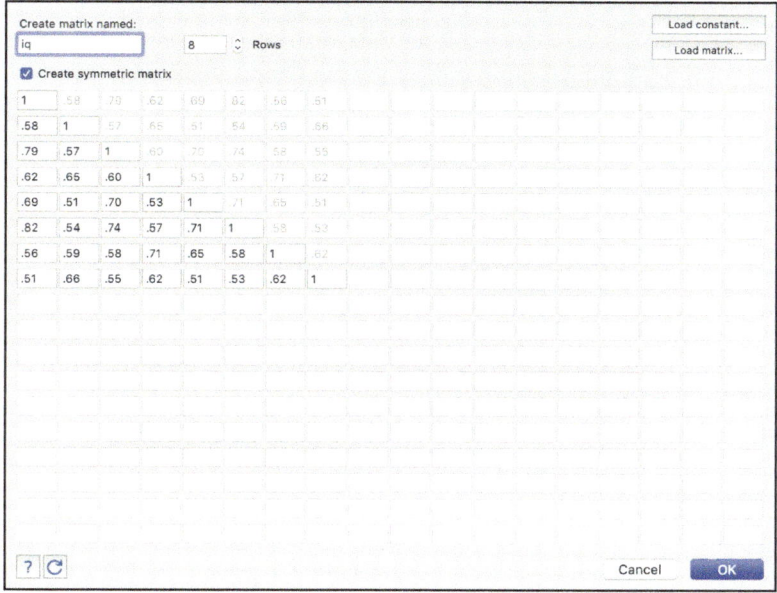

Figure 4.7 Stata input matrix window

The variable names and number of participants can be entered on this window and the correlation matrix can be entered in a separate window generated by the ***Input matrix*** button (Figure 4.7). Giving the matrix a name (iq) and dimension (8) and identifying it as symmetric will allow manual entry of the correlation matrix.

Once the correlation matrix has been input and the **OK** button clicked, Stata returns to the **factormat – Factor analysis of a matrix** window (Figure 4.6) where the name of the matrix can be specified and the factor analysis subsequently conducted via the ***Model 2*** button.

A final option for correlations if raw data is used as input is to use the user-contributed makematrix module that can be found via the search makematrix command.

Saving Data

It might be useful to save the imported iq data as a native Stata data file so that the importation process will not have to be repeated. That is accomplished via ***File > Save As***. Provide a name (iq) and disk location to complete this operation. Stata will automatically recognize that data are being saved and will use the proper file format (.dta). Alternatively, the data could have been exported in Excel, text, SAS, Stata, etc. format via ***File > Export***. To open the new Stata data file, simply click the **Open** icon shortcut or the ***File > Open*** menus and navigate to the iq.dta file.

Data Management

Stata includes a robust ensemble of data management options that are directly available through commands or via the **Data** and **View** menus. Mitchell (2020) provided detailed guidance on data management.

5 Decision Steps in Exploratory Factor Analysis

Researchers must make several thoughtful and evidence-based methodological decisions while conducting an EFA (Henson & Roberts, 2006). There are a number of options available for each decision, some better than others (Lloret et al., 2017).

Those decisions (steps in the EFA process) are depicted in Figure 5.1. Note that each decision might be sufficient for movement to the next step or it might necessitate a return to a previous step. This visual presentation emphasizes the exploratory nature of EFA and the knowledge that the evidence for some decisions is uncertain or dependent upon prior results.

There are numerous decision steps and each step contains several complex components. Accordingly, users are encouraged to print Figure 5.2 and use it as a checklist as they conduct an EFA. That practice will ensure consistent implication of evidence-based, best practice decisions (Watkins, 2009).

DOI: 10.4324/9781003149286-5

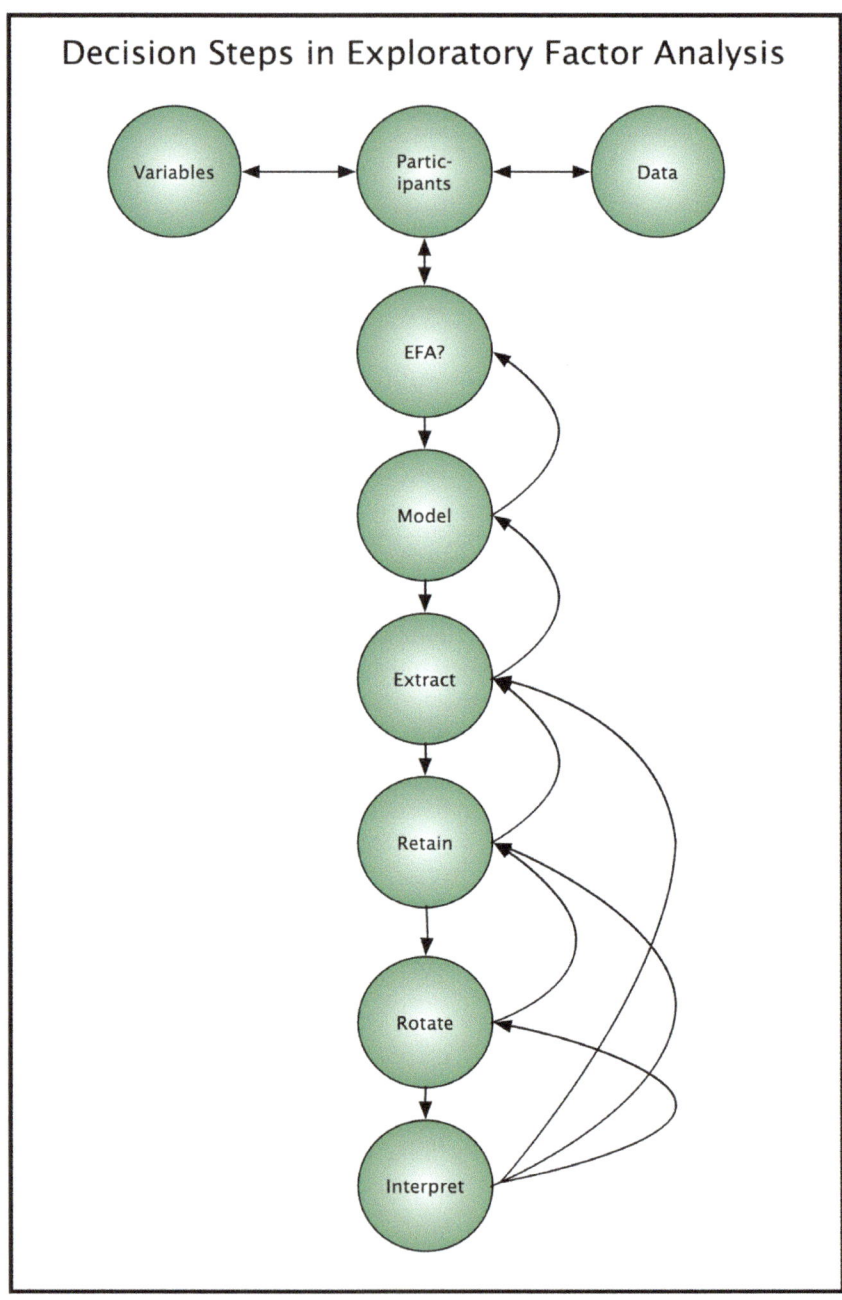

Figure 5.1 Flow chart of decision steps in exploratory factor analysis

Decision Steps in Exploratory Factor Analysis

What variables to include
- Number of variables per factor
- Adequate representation of the domain
- Avoid low communality
- Avoid low reliability
- Variables cannot be dependent upon each other

What participants to include
- Number of participants
- Adequate representation of the population

Is data appropriate
- Accuracy (out of range values, plausible summary values)
- Missing data (amount and distribution)
- Univariate and multivariate outliers
- Linearity
- Univariate and multivariate normality

Is EFA appropriate
- Bartlett's test of sphericity
- Kaiser-Meyer-Olkin test of sampling adequacy
- Correlation matrix

Model of factor analysis
- Principal components analysis
- Common factor analysis

Factor extraction method
- Weak factors/Nonnormal: Least-squares, Principal Axis
- Multivariate normal: Maximum Likelihood

How many factors to retain
- Parallel analysis
- Minimum average partials (MAP)
- Visual scree
- Theoretical convergence and parsimony
- Others

Rotate factors
- Orthogonal: Varimax
- Oblique: Oblimin, Promax

Interpret results
- Simple structure
- Theoretical convergence and parsimony

Report results
- All decision steps

Figure 5.2 Checklist of decision steps in exploratory factor analysis

6   Step 1

Variables to Include

The choice of variables to include in an exploratory factor analysis (EFA) is important because "if the indicators are not selected well, the recovered structure will be misleading and biased" (Little et al., 1999, p. 209). Consequently, the measured variables must be selected after careful consideration of the domain of interest (Cattell, 1978; Widaman, 2012). Some domains will be relatively narrow (e.g., depression) whereas others will be broad (e.g., psychopathology). Nevertheless, variables that assess all important aspects of the domain of interest should be sampled (Carroll, 1985; Wegener & Fabrigar, 2000).

Psychometric Properties

The psychometric properties of the measured variables must also be carefully considered. When EFA is conducted on measured variables with low communalities (those that do not share much common variance), substantial distortion can result. One obvious reason for low communality is poor reliability. Scores can vary due to true responses of the examinees or due to error. The reliability coefficient estimates the proportion of true score variance. For example, a reliability coefficient of .40 indicates 40% true score variance and 60% error. Error variance, by definition, cannot be explained by factors. Because of this, variables with low reliability will have little in common with other variables and should be avoided in EFA (Fabrigar et al., 1999). Therefore, the reliability of measured variables should be considered when selecting them for inclusion in an EFA (Feldt & Brennan, 1993; Watkins, 2017). "If $r_{xx} < .50$, then most of the total variance is due to measurement error. Indicators with such low score reliabilities should be excluded from the analysis" (Kline, 2013, p. 173). Nevertheless, decisions about variables to include in EFA must also consider the possibility that systematic error might have artificially inflated reliability estimates at the expense of validity (see Clifton, 2020 for a discussion of this phenomenon).

A second psychometric reason that a variable might have a low communality is that it is reliable, but unrelated to the domain of interest and thus shares little common variance with the variables that tap that domain. Thus, the validity of measured variables must also be respected (Messick, 1995).

DOI: 10.4324/9781003149286-6

This is related to reliability in that reliability is necessary but not sufficient for validity. Validity suggests that the relationship between the measured variables and factors should be congruent. Given that EFA assumes that the factors influence the measured variables, what measured variables make sense for the domain of interest? For instance, if the domain of interest is assumed to be depression, it makes no sense to include variables that measure body mass, family income, and height because they are not reasonably influenced by depression. This is an example of construct irrelevance. Likewise, it makes little sense to include only addition and subtraction problems as variables if the domain of interest is arithmetic. That domain includes multiplication and division as well as addition and subtraction. This is an example of construct underrepresentation (Spurgeon, 2017). Readers should consult texts on measurement and scale development to gain a better understanding of these psychometric issues (Bandalos, 2018; Clifton, 2020; Cooper, 2019; DeVellis, 2017; Kline, 2000).

Marker Variables

It might be useful to include variables with known properties that have previously been studied (marker variables) in an EFA if the remainder of the measured variables are relatively unknown (Carroll, 1985; Comrey & Lee, 1992; Gorsuch, 1988; Nunnally & Bernstein, 1994; Tabachnick & Fidell, 2019; Zhang & Preacher, 2015). For example, vocabulary tests have long been used as measures of verbal ability so a vocabulary test might be included in an EFA if several new tests that purport to measure verbal ability are analyzed.

Formative versus Effect Indicators

In theory, there are situations where the measured variables are more properly treated as determinants rather than effects of latent variables (Edwards & Bagozzi, 2000). These are called *formative* indicators. That is, the causal direction goes from measured variables to the latent variable. For example, loss of job and divorce are measures of exposure to stress that might best be thought of as causal indicators. Each event creates stress rather than the reverse. Likewise, education, income, and occupational prestige may be causal indicators of socioeconomic status (SES). Clearly, more education and income and a high prestige job cause higher SES; SES does not cause these indicators. Thus, losing one's job would lead to lower SES, but this lowered status would leave one's years of education unchanged. Eliminating one causal indicator (say, education) from the model changes the meaning of the SES scale because part of the SES construct is not represented. For most purposes, we will assume *effect* indicators and proceed accordingly but the type of indicators should be considered because application of EFA with formative indicators may produce severely biased results (Rhemtulla et al., 2020).

Number of Variables

Two measured variables are sometimes used to identify a factor. However, at least three variables are necessary for statistical identification (Goldberg & Velicer, 2006; Kano, 1997; Mulaik, 2010; Streiner, 1994). Imagine that the sphere pictured in Figure 6.1 represents multidimensional space. The star is the construct. If there are only two variables (A and B), drawing a line between them (to simplistically represent a perfect linear relationship) does not encompass the construct. However, adding variable C and its relationship with the other measured variables does enclose the construct. Factors formed by only two variables are underdetermined and may be unstable and cause improper solutions or nonconvergence issues (Goldberg & Velicer, 2006; Hahs-Vaughn, 2017; Kano, 1997).

Given this information, how many variables should be included in an EFA? Too few variables per factor "becomes not only an issue of model identification and replication, but also a matter of construct underrepresentation" (Schmitt et al., 2018, p. 350). The number of measured variables will also impact the sample size decision. The requisite number of variables is also influenced by the reliability and validity of the measured variables. Fewer variables might be needed if they exhibit high reliability and validity. For example, subscales from individually administered tests of cognitive ability typically meet this definition and only three to four per factor might be needed. It is generally recommended that at least three or four to six reliable variables representing each common factor be included in an analysis (Carroll, 1985; Comrey & Lee, 1992; Fabrigar & Wegener, 2012; Fabrigar

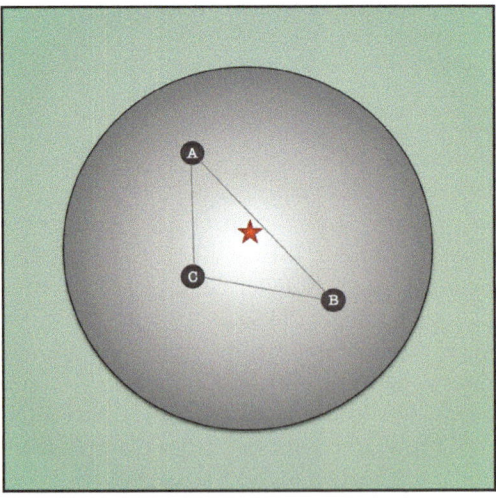

Figure 6.1 Variables in multidimensional space

et al., 1999; Goldberg & Velicer, 2006; Goretzko et al., 2019; Gorsuch, 1988; Hair et al., 2019; Hancock & Schoonen, 2015; Kline, 2013; McCoach et al., 2013; McDonald, 1985; Mulaik, 2010; Streiner, 1994; Tabachnick & Fidell, 2019). In contrast, individual items on both ability and personality tests are generally of low reliability so many more might be needed to adequately represent the domain of interest. Kline (1994) recommended that at least ten items should be included for finalizing a test and many more if the test is in the development stage.

Statistical Issues

There are several situations where the choice of variables can create statistical problems that preclude EFA. First, if there are more variables than participants. Second, if the number of participants with scores on each variable are not equal. This could be caused by pairwise deletion of missing data. Third, if there are linear dependencies among the variables. For example, variables that are composites of other variables. Singularity is the term used when measured variables are linearly dependent on each other. Other examples of inappropriate variables are "right" and "wrong" and ipsative scores. Statistically, singularity prevents the matrix operations needed for EFA to be performed (Pett et al., 2003). Finally, EFA is not possible if a variable exhibits zero variance.

Short of singularity, a statistical problem akin to dividing by zero in arithmetic can appear with extreme multicollinearity; that is, when correlations among the measured variables are too high (Lorenzo-Seva & Ferrando, 2021). Tabachnick and Fidell (2019) suggested that correlations above .90 among measured variables might indicate extreme multicollinearity. Statistically, multicollinearity causes the matrix operations needed for EFA to produce unstable results (Pett et al., 2003; Rockwell, 1975).

Technically, these situations can cause the statistical analysis to produce the equivalent of a negative variance called a nonpositive definite matrix or communality estimates greater than one that are called Heywood cases (Flora, 2018). These are mathematically improper solutions (van Driel, 1978) that "should be discarded and not interpreted" (Flora, 2018, p. 257). Detailed descriptions of these issues have been provided by Lorenzo-Seva and Ferrando (2021), van Driel (1978), and Wothke (1993).

Report

The eight measured variables in this study were developed to measure cognitive ability. Based on typical ability tests, these variables are expected to exhibit good psychometric properties. As suggested by Fabrigar et al. (1999), four variables were hypothesized to represent each factor.

7 Step 2

Participants

Characteristics of Participants

First, which participants? This is primarily a matter of logic and common sense. To which population are the results to generalize? Sample from that population. Does the sample make sense given the factors you are attempting to measure?

Child (2006) warned against using samples collected from different populations to compute correlations because factors that are specific to a population might be obscured when pooled. Tabachnick and Fidell (2019) and Comrey and Lee (1992) also warned about pooling the results of several samples, or the same sample measured across time. Sampling procedures are acceptable, but there must be a clear understanding of sample–population differences (Widaman, 2012).

Additionally, negligent responses from unmotivated participants may introduce bias. Woods (2006) found that factor analysis results were affected when more than 10% of the participants responded carelessly. Thus, the validity of participants' responses must be considered.

Number of Participants

Beyond which participants to include in an EFA, it is also important to know how many participants to include. Correlation coefficients tend to be less stable when estimated from small samples. For example, with a true population correlation of zero and a sample size of 100, about 95% of the correlations will fall between −.20 and +.20. In contrast, about 95% of the correlations will fall between −.09 and +.09 when the sample size is 500. One simulation study found that 1,000 participants were needed to estimate correlation coefficients within ± .05 and 250 participants were needed for "reasonable trade-offs between accuracy and confidence" (Schönbrodt & Perugini, 2013, p. 611).

Guidelines for estimating the number of participants required for an EFA have focused on the: (a) absolute number of participants, (b) ratio of participants to measured variables, (c) quality of the measured variables, and (d) ratio of measured variables to factors. In the first case, Comrey and Lee

DOI: 10.4324/9781003149286-7

(1992) suggested that 100 participants is poor, 200 is fair, 300 is good, 500 is very good, and 1,000 or more is excellent. Regarding the participant to variable ratio, both Child (2006) and Gorsuch (1983) recommended five participants per measured variable with a minimum of 100 participants. Other measurement experts have suggested a 10:1 or 20:1 ratio of participants to measured variables (Benson & Nasser, 1998; Hair et al., 2019; Osborne & Banjanovic, 2016). Unfortunately, "these guidelines all share three characteristics: (1) no agreement among different authorities, (2) no rigorous theoretical basis provided, and (3) no empirical basis for the rules" (Velicer & Fava, 1998, p. 232).

In contrast, sample size guidelines based on the quality of measured variables and the variable to factor ratio have often been based on statistical simulation studies where the number of variables, number of participants, number of variables per factor (factor overdetermination), and the percent of variance accounted for by the factors (communality) were systematically modified and evaluated (Guadagnoli & Velicer, 1988; Hogarty et al., 2005; MacCallum et al., 2001; MacCallum et al., 1999; Mundfrom et al., 2005; Velicer & Fava, 1998; Wolf et al., 2013). These simulation studies have used dissimilar numbers of factors, variables, variance estimates, EFA methods, etc. so their results are not always compatible. However, factor overdetermination (i.e., number of measured variables per factor) and communality were consistently found to be important determinants of sample size. Greater factor overdetermination and larger communality tended to require smaller sample sizes than the reverse. Further, sample size, factor overdetermination, and communality seemed to interact so that "strength on one of these variables could compensate for a weakness on another" (Velicer & Fava, 1998, p. 243). Given these results, try to include variables with high communalities (≥ .60; Gibson et al., 2020), overdetermined factors (> 3 variables per factor), and a parsimonious number of factors.

Although "it is impossible to derive a minimum sample size that is appropriate in all situations" (Reise et al., 2000, p. 290), the subjective guidelines for good factor recovery enumerated in Figure 7.1 are consistent with the tables provided by Mundfrom et al. (2005) and the results reported in other studies (Gorsuch, 1983; Guadagnoli & Velicer, 1988; Hogarty et al., 2005; MacCallum et al., 1999, 2001; Velicer & Fava, 1998). However, "factor analysis should be viewed as inherently a subject-intensive enterprise" (Goldberg & Velicer, 2006, p. 214) so "larger samples are better than smaller samples" (Osborne et al., 2007, p. 90).

Two additional considerations in determining the appropriate sample size are type of data and amount of missing data. The recommendations in Figure 7.1 are based on data from continuous, normally distributed data. Dichotomous variables (two response options) are neither continuous nor normally distributed and will require three to ten times more participants (Pearson & Mundfrom, 2010; Rouquette & Falissard, 2011).

For ordinal data, a simulation study by Rouquette and Falissard (2011) suggested that stable and accurate factor solutions can be obtained with: (a)

Communality	Variables per Factor	Number of Factors	Number of Participants
	3	2	100
	3	3	170
	3	4	260
	3	5	300
	4	2	100
	4	3	120
≥ .60	4	4	170
	4	5	220
	5	3	100
	5	4	100
	5	5	130
	10	6	100
	3	2	160
	3	3	450
	3	4	500
	3	5	700
	4	3	130
.20 − .80	4	4	240
	4	5	320
	5	3	100
	5	4	110
	5	5	140
	10	6	100
	3	3	1,200
	3	4	1,200
	3	5	1,300
	4	3	230
≤ .20	4	4	250
	4	5	400
	5	3	150
	5	4	170
	5	5	180
	10	6	150

Figure 7.1 Sample size estimates based on communality and variable:factor ratio

350 to 400 participants when there are ten variables per factor; (b) 400 to 450 participants when there are seven or eight variables per factor; and (c) 500 to 600 participants when there are five or six variables per factor.

Additionally, models with 10% missing data may require a 25% increase in sample size and models with 20% missing data may need a 50% increase in sample size (Wolf et al., 2013). Regardless of type of data and missing values, more data is always better "because the probability of error is less, population estimates are more accurate, and the findings are more generalizable" (Reio & Shuck, 2015, p. 15).

Report

The participants in this study were children in grades two through six who were referred for assessment for consideration of special educational accommodations. The average communality of variables similar to those in this study is ≥ .60 (Dombrowski et al., 2018). Additionally, these cognitive variables were relatively reliable (α ≥ .80) and normally distributed. Given the communality, number of factors, and number of variables, 100 participants would be needed for good factor recovery (Mundfrom et al., 2005). Thus, the current sample of 152 participants appears to be adequate for EFA.

8 Step 3

Data Screening

Effective data screening involves inspection of both statistics and graphics (Hoelzle & Meyer, 2013; Malone & Lubansky, 2012). Either alone is insufficient. This was famously demonstrated by Anscombe (1973) who created four x–y data sets with relatively equivalent summary statistics as displayed in Figure 8.1.

```
 variable |        N      mean        sd       min       max|
----------+----------------------------------------------------
       x1 |       11         9  3.316625         4        14
       y1 |       11  7.500909  2.031568      4.26     10.84
       x2 |       11         9  3.316625         4        14
       y2 |       11  7.500909  2.031657       3.1      9.26
       x3 |       11         9  3.316625         4        14
       y3 |       11       7.5  2.030424      5.39     12.74
       x4 |       11         9  3.316625         8        19
       y4 |       11  7.500909  2.030579      5.25      12.5
----------------------------------------------------------------


          |       x1        y1        x2        y2        x3        y3        x4        y4
----------+--------------------------------------------------------------------------------
       x1 |   1.0000
       y1 |   0.8164    1.0000
       x2 |   1.0000    0.8164    1.0000
       y2 |   0.8162    0.7500    0.8162    1.0000
       x3 |   1.0000    0.8164    1.0000    0.8162    1.0000
       y3 |   0.8163    0.4687    0.8163    0.5879    0.8163    1.0000
       x4 |  -0.5000   -0.5291   -0.5000   -0.7184   -0.5000   -0.3447    1.0000
       y4 |  -0.3140   -0.4891   -0.3140   -0.4781   -0.3140   -0.1555    0.8165    1.0000
```

Figure 8.1 Descriptive statistics and correlation matrix for Anscombe quartet data

However, there is danger in relying on summary statistics alone. When this "Anscombe quartet" is graphed, as in Figure 8.2, the real relationships in the data emerge. Specifically, the x1–y1 data appear to follow a rough linear relationship with some variability, the x2–y2 data display a curvilinear rather than a linear relationship, the x3–y3 data depict a linear relationship except for one large outlier, and the x4–y4 data show x remaining constant except for one (off the chart) outlier.

Assumptions

All multivariate statistics are based on assumptions that may bias results if they are violated. The assumptions of exploratory factor analysis (EFA) are mostly

DOI: 10.4324/9781003149286-8

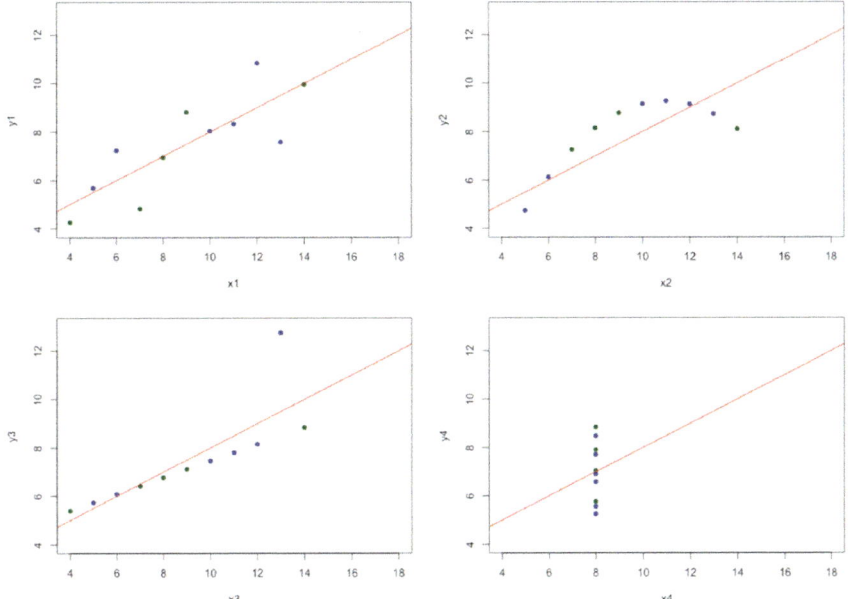

Figure 8.2 Scatterplots for Anscombe quartet data

conceptual: It is assumed that some underlying structure exists, that the relationship between measured variables and the underlying common factors are linear, and that those linear relationships are invariant across participants (Fabrigar & Wegener, 2012; Hair et al., 2019).

However, EFA is dependent on Pearson product–moment correlations that also make statistical assumptions. Specifically, it is assumed that a linear relationship exists between the variables and that there is an underlying bivariate normal distribution (Sheskin, 2011). To meet these assumptions, variables must be measured on a continuous scale (Bandalos, 2018; Puth et al., 2015; Walsh, 1996). Violation of the assumptions that underlie the Pearson product–moment correlation may bias EFA results. As suggested by Carroll (1961), "there is no particular point in making a factor analysis of a matrix of raw correlation coefficients when these coefficients represent manifest relationships which mask and distort latent relationships" (p. 356).

More broadly, anything that influences the correlation matrix can potentially affect EFA results (Carroll, 1985; Flora et al., 2012; Onwuegbuzie & Daniel, 2002). As noted by Warner (2007), "because the input to factor analysis is a matrix of correlations, any problems that make Pearson *r* misleading as a description of the strength of the relationship between pairs of variables will also lead to problems in factor analysis" (p. 765). Accordingly, the data must be screened before conducting an EFA to ensure that some untoward influence has not biased the results (Flora et al., 2012; Goodwin &

Leech, 2006; Hair et al., 2019; Walsh, 1996). Potential influences include restricted score range, linearity, data distributions, outliers, and missing data. "Consideration and resolution of these issues before the main analysis are fundamental to an honest analysis of the data" (Tabachnick & Fidell, 2019, p. 52).

Stata offers numerous routines to explore data. For example, the codebook and describe commands will display descriptions of each variable. Likewise, the summarize and correlate commands will produce descriptive statistics and correlations, respectively. Descriptive statistics can also be obtained through the menu system by selecting *Data > Describe data > Describe data contents (codebook)* or *Data > Describe data > Summary statistics* options. A correlation matrix can also be generated through the menu system by selecting *Statistics > Summaries, tables and tests > Summary and descriptive statistics > Correlations and covariances* options.

It may be instructive to manually generate and examine output. Thus, each aspect of the data that might affect EFA results is subsequently explored. The extensive roster of statistical procedures available in Stata are displayed in Figure 8.3.

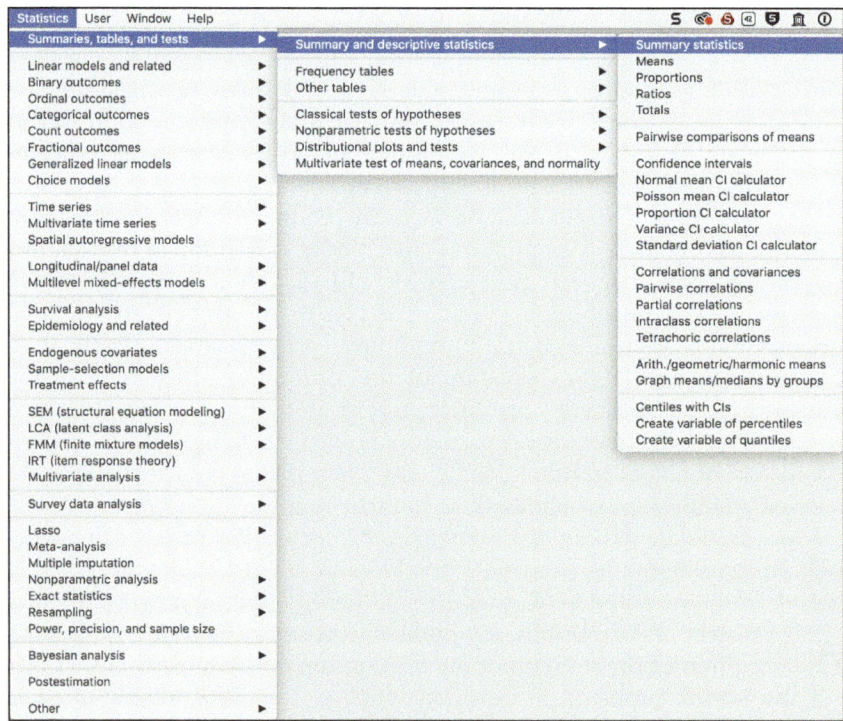

Figure 8.3 Stata statistics menu system

Restricted Score Range

The range of scores on the measured variables must be considered. If the sample is more homogeneous than the population, restriction of range in the measured variables can result and thereby attenuate correlations among the variables. At the extreme, a measured variable with zero variance will cause statistical estimation problems and inadmissible EFA results (Lorenzo–Seva & Ferrando, 2021; Wothke, 1993). Less extreme attenuation can result in biased EFA estimates. For example, using quantitative and verbal test scores from the 1949 applicant pool of the U.S. Coast Guard Academy, the quantitative and verbal test score correlations dropped from .50 for all 2,253 applicants to only .12 for the 128 students who entered the Academy (French et al., 1952). In such cases, a "factor cannot emerge with any clarity" (Kline, 1991, p. 16).

A compact display of descriptive statistics can be examined to evaluate restriction of range among these iq variables. Figure 8.4 contains that display as well as the `tabstat` command that generated it. Alternative commands include `summarize` and `codebook`, `compact`. The population standard deviation for these variables is 15 and the sample standard deviations range from 13.9 to 17.3. Thus, restriction of range does not appear to be a problem.

```
.tabstat vocab1-designs2, statistics(count mean sd min max skewness kurtosis) columns (statistics)

    variable |      N        mean         sd      min      max    skewness   kurtosis
-------------+-------------------------------------------------------------------------
      vocab1 |    152        97.5    17.34152       55      137      -.3148     2.9938
    designs1 |    152    97.65132     14.4732       58      130      -.2126     2.9807
    similar1 |    152    103.5921    17.25552       55      145      -.4533     3.3075
     matrix1 |    152    99.53289    16.60564       55      134      -.5239     3.1140
    veranal2 |    152    101.5066    14.76886       57      134      -.6227     3.3883
      vocab2 |    152    100.6316    16.41623       37      144      -.8117     4.7875
     matrix2 |    152    101.4474      16.172       49      137      -.5740     3.8318
    designs2 |    152    100.6447     13.9224       45      137      -.4537     4.3583
```

Figure 8.4 Compact tabstat display of descriptive statistics

As displayed in Figure 8.5, a correlation matrix (lower diagonal) can be generated by the `correlate` command. Note that `correlate` assumes that all the variables are to be included in the correlation matrix, identical to the `correlate vocab1-designs2` command.

```
             | vocab1 designs1 similar1  matrix1 veranal2   vocab2  matrix2 designs2
-------------+-------------------------------------------------------------------------
      vocab1 | 1.0000
    designs1 | 0.5762   1.0000
    similar1 | 0.7885   0.5697   1.0000
     matrix1 | 0.6179   0.6500   0.5985   1.0000
    veranal2 | 0.6892   0.5086   0.7015   0.5328   1.0000
      vocab2 | 0.8242   0.5390   0.7353   0.5659   0.7059   1.0000
     matrix2 | 0.5637   0.5862   0.5818   0.7146   0.6480   0.5773   1.0000
    designs2 | 0.5074   0.6616   0.5505   0.6190   0.5137   0.5303   0.6244   1.0000
```

Figure 8.5 Correlation matrix from correlate command

The correlate command will default to listwise deletion for missing data. In contrast, the pwcorr command applies pairwise deletion for missing data. As described in the *Variables to Include* chapter, pairwise deletion might create statistical problems for EFA and should be avoided.

Linearity

Pearson coefficients are measures of the linear relationship between two variables. That is, their relationship is best approximated by a straight line. Curvilinear or nonlinear relationships will not be accurately estimated by Pearson coefficients. Although subjective, visual inspection of scatterplots can be used to assess linearity.

The extensive graphical procedures available in Stata are displayed in Figure 8.6. Scatterplots can be obtained via the ***Twoway graph (scatter, line, etc.)*** menu option that opens a new window where the ***Create*** option is selected that, in turn, opens a new window where the type of graph (***Scatter***) is selected and the two variables to plot are identified. After accepting these options, the ***Plot definitions*** box will display the title of the new plot that was defined (Plot 1).

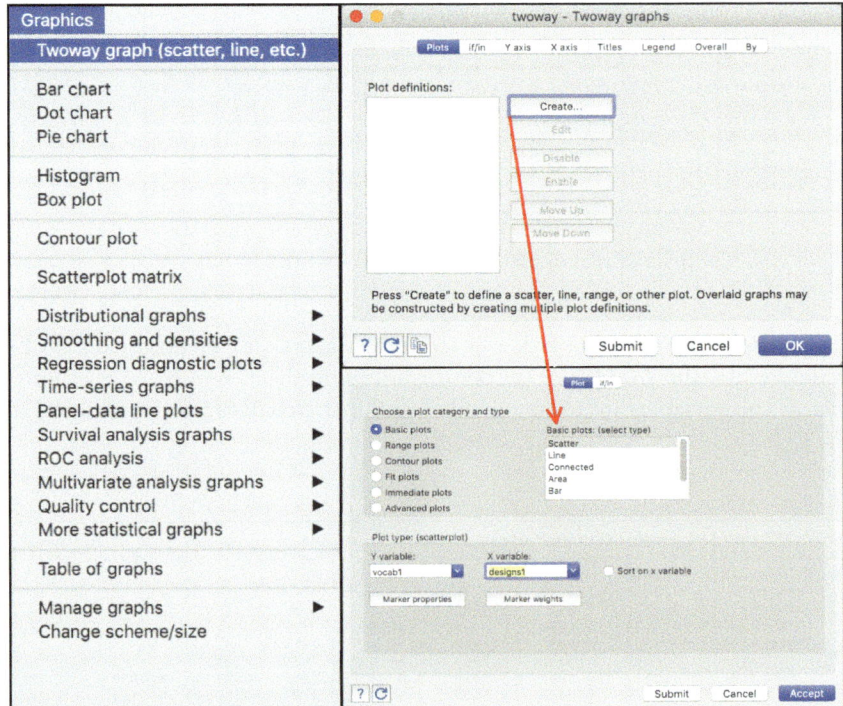

Figure 8.6 Stata graphics menu system

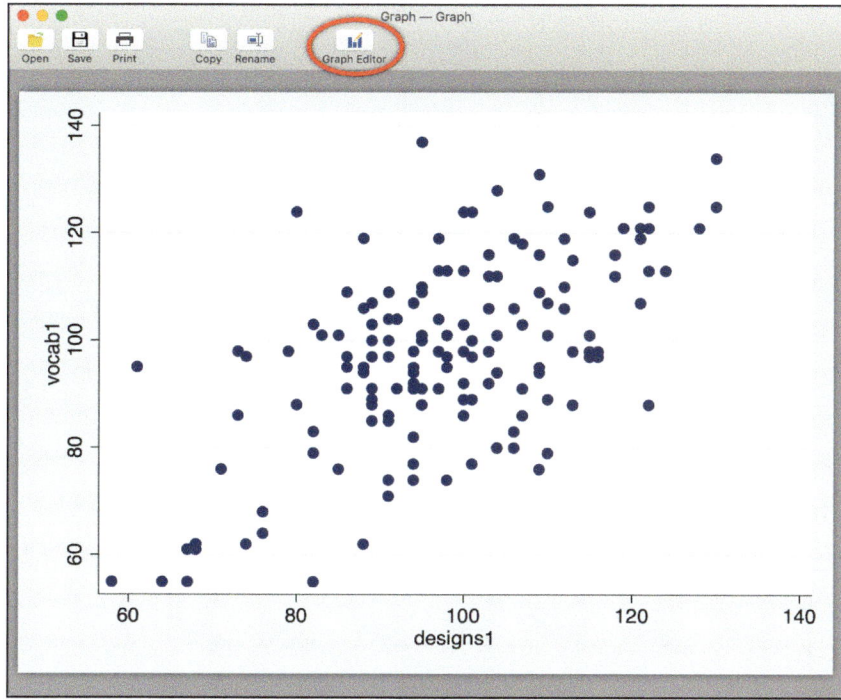

Figure 8.7 Scatterplot of two variables in graph window

Clicking the **OK** button opens a new Graph window that displays the scatterplot of the vocab1 and designs1 variables (Figure 8.7). A command to produce this scatterplot is much simpler: twoway (scatter vocab1 designs1). A scatterplot with a superimposed linear best fit line can be produced with the twoway (lfit vocab1 designs1) (scatter vocab1 designs1) command and a 95% confidence band for the best fit line can be included with the twoway (lfitci vocab1 designs1) (scatter vocab1 designs1) command. Every property of the scatter plot can be edited by clicking the **Graph Editor** shortcut icon. Alternatively, Graphs can be saved in Stata's graphics format (.gph) or in a generic graphics format (.eps, .gif, .jpeg, .pdf, .png, .svg, or .tiff) by clicking the **Save** shortcut icon or by selecting the **File > Save As** menu options.

It may be more efficient to review a scatterplot matrix that contains all of the variables rather than individual scatterplots for each pair of variables. **Scatterplot matrix** is one option under the **Graphics** menu. Alternatively, the command graph matrix vocab1-designs2 will produce a scatterplot matrix (Figure 8.8). As with the scatterplot, every property of this graph can be edited by clicking the **Graph Editor** shortcut icon. The scatterplot matrix can be saved in Stata's graphics format (.gph) or in a generic graphics format (.eps, .gif, .jpeg, .pdf, .png, .svg, or .tiff) by clicking the **Save** shortcut icon or by selecting the **File > Save As** menu options.

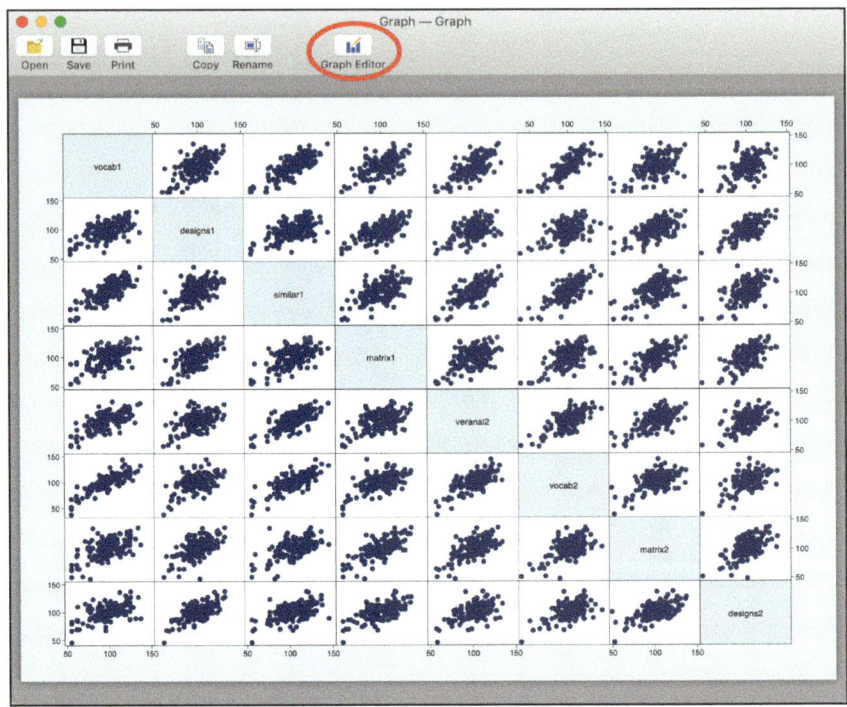

Figure 8.8 Scatterplot matrix in graph window

Graph properties can also be specified in the `graph` command. For example, the `graph matrix vocab1-designs2, mcolor(maroon) msymbol(diamond)    msize(3-pt)    scheme(s1color)` command will produce a scatterplot matrix (Figure 8.9) with modified color and size properties. That graph can, in turn, be edited in the ***Graph Editor***.

After reviewing the scatterplots, it appears that the iq variables are linearly related.

Data Distributions

Pearson correlation coefficients theoretically range from -1.00 to $+1.00$. However, that is only possible when the two variables have the same distribution. If, for example, one variable is normally distributed and the other distribution is skewed, the maximum value of the Pearson correlation is less than 1.00. The more the distribution shapes differ, the greater the restriction of r. Consequently, it is important to understand the distributional characteristics of the measured variables included in an EFA. For example, it has long been known that dichotomous items that are skewed in opposite directions may produce what are know as difficulty factors when submitted to EFA (Bernstein & Teng, 1989; Greer et al., 2006). That is, a factor may appear that is an artifact of variable distributions rather than the effect of their content.

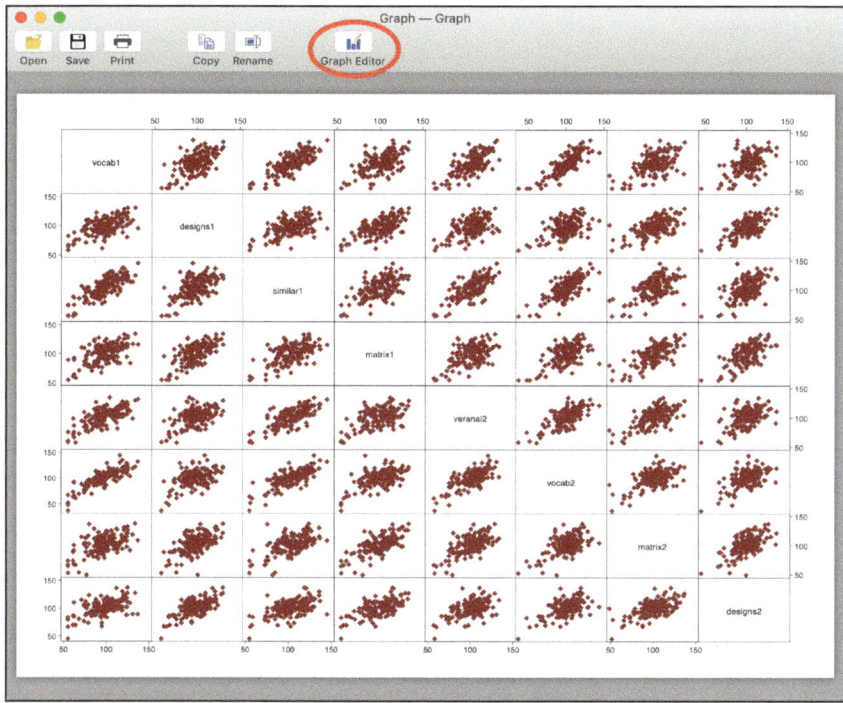

Figure 8.9 Edited scatterplot matrix in graph window

As demonstrated in Figure 8.4, a compact and informative table of descriptive statistics can be generated by the command `tabstat vocab1-designs2, statistics(count mean sd min max skewness kurtosis) columns(statistics)`. A long-form descriptive statistics display can be created with the `summarize vocab1-designs2, detail` command. Ordinarily, a normal distribution will have skew = 0 and kurtosis = 0. Using this standard, skew > 2.0 or kurtosis > 7.0 would indicate severe univariate nonnormality (Curran et al., 1996). However, in Stata a normal distribution will have skew = 0 and kurtosis = 3. Thus, three can be subtracted from Stata's kurtosis values to arrive at the more commonly reported kurtosis statistic. A user-contributed ado-file can be downloaded via `ssc install moments2` that will compute zero-normed kurtosis values (Figure 8.10).

The univariate statistics in Figure 8.4 indicate that all eight measured variables are relatively normally distributed (skew < 1.0 and zero-normed kurtosis < 2.0) so there should be little concern about correlations being restricted due to variable distributions. Skew (departures from symmetry) and kurtosis (distributions with heavier or lighter tails and higher or flatter peaks) of all variables seem to be close to normal.

Graphs can be useful for visual verification of this conclusion. A boxplot that displays the distributional statistics of the measured variables can be

```
. moments2
--------------------------------------------------------------------
   n = 152 |       mean           SD      skewness      kurtosis
-----------+--------------------------------------------------------
    vocab1 |     97.500       17.342        -0.318         0.034
  designs1 |     97.651       14.473        -0.215         0.021
  similar1 |    103.592       17.256        -0.458         0.358
   matrix1 |     99.533       16.606        -0.529         0.158
  veranal2 |    101.507       14.769        -0.629         0.442
    vocab2 |    100.632       16.416        -0.820         1.888
   matrix2 |    101.447       16.172        -0.580         0.900
  designs2 |    100.645       13.922        -0.458         1.445
--------------------------------------------------------------------
```

Figure 8.10 Descriptive statistics from moments2 ado-file

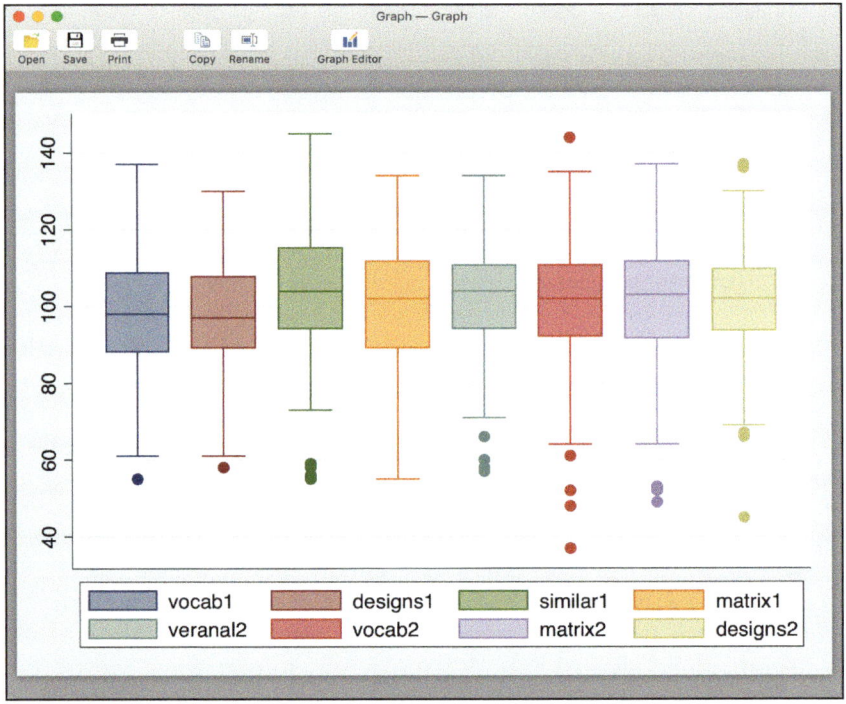

Figure 8.11 Boxplot for iq data

generated via **Graphics > Box plot > vocab1–designs2** or the graph box vocab1-designs2 command (Figure 8.11). Boxplots have the following characteristics: the thick line in the box is the median, the bottom of the box is the first quartile (25th percentile), the top of the box is the third quartile (75th percentile), the "whiskers" show the range of the data (excluding outliers), and the circles identify outliers (defined as any value 1.5 times the interquartile range).

Although there are a few data points outside the 1.5 interquartile range, the median of each variable appears roughly centered in the variable boxes. Combined with the scatterplot, this boxplot reinforces the conclusion of univariate normality drawn from the descriptive statistics.

A group of measured variables might exhibit univariate normality and yet be multivariate nonnormal (DeCarlo, 1997). That is, the joint distribution of all the variables might be nonnormal. Stata offers a command that computes several measures of multivariate normality: `mvtest normality vocab1-designs2, stats(all)`. Of these, Mardia's (1970) measures are probably the most widely accepted (Cain et al., 2017) and Mardia's kurtosis is typically used to evaluate the distribution (Finney & DiStefano, 2013). The expected value of Mardia's kurtosis for a multivariate normal population is $v(v + 2)$ where v is the number of variables. Thus, the expected kurtosis is 80 for the iq data and, as shown in Figure 8.12, multivariate kurtosis was 84.12 ($p < .05$). Nonnormality, especially kurtosis, can bias Pearson correlation estimates and thereby bias EFA results (Curran et al., 1996; Flora et al., 2012; Pett et al., 2003; Sheskin, 2011; Yong & Pearce, 2013).

```
. mvtest normality vocab1-designs2, stats(all)

Test for multivariate normality
Mardia mSkewness =   6.275617    chi2(120) =  162.830    Prob>chi2 =  0.0057
Mardia mKurtosis =  84.12349     chi2(1) =      4.038    Prob>chi2 =  0.0445
Henze-Zirkler    =   .9824914    chi2(1) =      0.903    Prob>chi2 =  0.3419
Doornik-Hansen                   chi2(16) =     32.322    Prob>chi2 =  0.0091
```

Figure 8.12 Multivariate normality measures via mvtest command

Alternatively, an online calculator is available at https://webpower. psychstat.org/models/kurtosis/ that can accept an Excel data file as input.

The extent to which variables can be nonnormal and not substantially affect EFA results has been addressed by several researchers. Curran et al. (1996) opined that univariate skew should not exceed 2.0 and univariate kurtosis should not exceed 7.0. Other measurement specialists have agreed with those guidelines (Bandalos, 2018; Fabrigar et al., 1999; Wegener & Fabrigar, 2000). In terms of multinormality, statistically significant multivariate normalized kurtosis values > 3.0 to 5.0 might bias factor analysis results (Bentler, 2005; Finney & DiStefano, 2013; Mueller & Hancock, 2019). Spearman or other types of correlation coefficients might be more accurate in those instances (Bishara & Hittner, 2015; Onwuegbuzie & Daniel, 2002; Puth et al., 2015). Given these univariate guidelines, it seems unlikely that the distributional characteristics of the iq will bias Pearson correlation estimates. On the other hand, the multivariate kurtosis is slightly elevated (84.12 versus 80), but only marginally significant ($p = .045$), which is weak statistical evidence (Anderson, 2020) that does not approach the .005 threshold recommended by some statisticians (Benjamin & Berger, 2019).

Outliers

As described by Tabachnick and Fidell (2019), "an outlier is a case with such an extreme value on one variable (a univariate outlier) or such a strange combination of scores on two or more variables (multivariate outlier) that it distorts statistics" (p. 62). Outliers are, therefore, questionable members of the data set. Outliers may have been caused by data collection errors, data entry errors, a participant not understanding the instructions, a participant deliberately entering invalid responses, or a valid but extreme value. Not all outliers will influence the size of correlation coefficients and subsequent factor analysis results but some may have a major effect (Liu et al., 2012). For example, the correlation between the matrix1 and designs1 variables in the iq data set is .65. That correlation drops to .10 when the final value in the matrix1 variable was entered as −999 rather than the correct value of 80. A data point like this might be caused by a typographical error or by considering a missing data indicator to be a real data point.

Obviously, some outliers can be detected by reviewing descriptive statistics. The minimum and maximum values might reveal data that exceeds the possible values that the data can take. For example, it is known that the values of the iq variables can reasonably range from around 40 to 160. Any value outside that range is improbable and must be addressed. One way to address such illegal values is to replace them with a missing value indicator. In Stata, missing data can be indicated by any character(s) the user specifies. It will be important to select missing value indicators that are unlikely to represent real data points. For example, the iq data is known to vary from around 40 to around 160 so missing data could be represented by −999, a value that is impossible in the real data.

There are no missing data in the iq data. However, the minimum value of 37 for the vocab2 variable may be too low to be a valid entry. The boxplot in Figure 8.11 revealed that several cases had a score lower than the 1.5 interquartile range, but did not identify the cases with those low scores. To do so, an ado-file contributed by a user must be installed via `ssc install extremes`. This package will list extreme values for each variable. For this example, 2.2 times the interquartile range will be used as recommended by some experts (Streiner, 2018) with the command `extremes variable, iqr(2.2)`. The results of this procedure for each variable revealed that cases 94 and 121 had extreme scores of 48 and 37, respectively, on the vocab2 variable and case 121 had an extreme score of 45 on the designs2 variable. By viewing cases 94 and 121 in the **Data Editor** with the `browse` command, it was apparent that these extreme scores were consistent with each case's scores on other variables so the data will be left unchanged. Additionally, they are univariate outliers and EFA is a multivariate procedure that necessitates that the multidimensional position of each data point be considered.

The Mahalanobis distance (D^2) is a measure of the distance of each data point from the mean of all data points in multidimensional space. Higher D^2

values represent observations farther removed from the general distribution of observations in multidimensional space and are, therefore, potential multi-variate outliers.

Another ado-file can be used to compute D^2 values in Stata. The command `ssc install mahapick` will install that procedure. The `mahascore vocab1-designs2, gen(mahd) refmeans compute_invcovarmat` command will generate a new variable named `mahd` that contains the D^2 value for each case. An id variable can be created with the `gen case_id = _n` command, followed by sorting the D^2 values from low to high with the `sort mahd` command. The `browse` command allows that variable to be viewed as in Figure 8.13.

D^2 follows a chi-square distribution with degrees of freedom equal to the number of variables and can, therefore, be tested for statistical significance but "it is suggested that conservative levels of significance (e.g., .005 or .001) be used as the threshold value for designation as an outlier" (Hair et al., 2019, p. 89). Stata can compute the critical value of a chi-square distribution. Thus, the critical value for probability of .005 and 8 degrees of freedom is computed as 21.95 by `display invchi2tail(8,.005)`.

The two largest D^2 values exceeded that critical value so cases 142 and 121 are potential outliers. An examination of case 121 shows that it contains the previously identified aberrant value of 37 for the vocab2 variable. However,

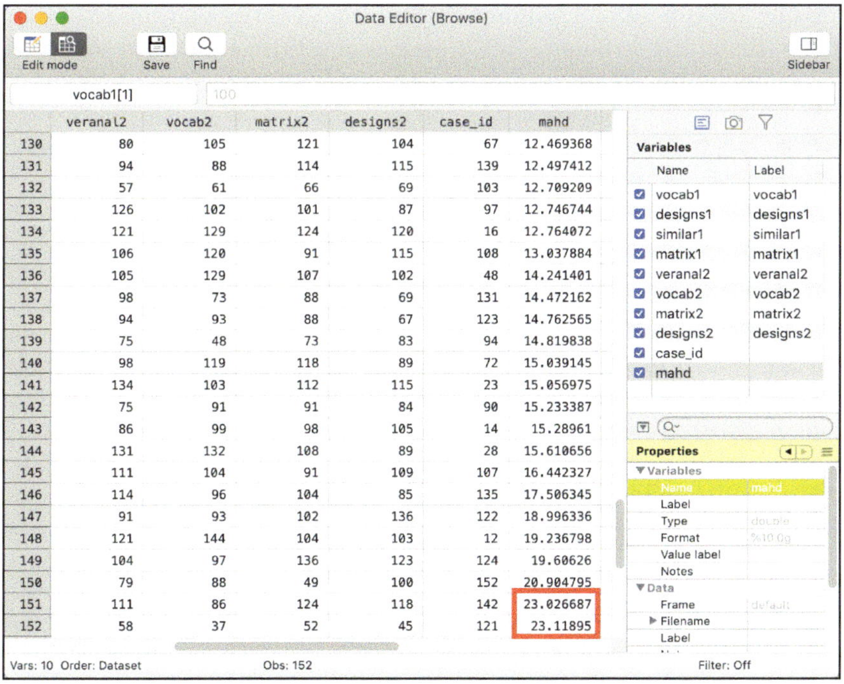

Figure 8.13 Data editor window with case id and Mahalanobis distance variables

all of the variable values for this case are very low (45 to 58) and consistent with impaired intellectual functioning. Given this consistency, there is no good reason to delete or modify the value of this case. Case 142 is not as easily understood. Some of its values are lower than average (e.g., 85) and others are higher than average (e.g., 127). There is no obvious explanation for why these values are discrepant but there is no reason to believe that it is not valid.

It is important to articulate an outlier policy prior to data analysis (Leys et al., 2018). Not to do so makes the researcher vulnerable to interpreting this ambiguous information inconsistent with best statistical practice (Simmons et al., 2011). Although there is considerable debate among statisticians as to the advisability of deleting outliers, Goodwin and Leech (2006) suggested that "the researcher should first check for data collection or data entry errors. If there were no errors of this type and there is no obvious explanation for the outlier—the outlier cannot be explained by a third variable affecting the person's score—the outlier should not be removed" (p. 260). Hair et al. (2019) expressed similar sentiments about outliers: "they should be retained unless demonstrable proof indicates that they are truly aberrant and not representative of any observations in the population" (p. 91). Alternative suggestions for identifying and reducing the effect of outliers have been offered (e.g., Tabachnick & Fidell, 2019). Regardless, extreme values might drastically influence EFA results so it is incumbent upon the researcher to perform a sensitivity analysis. That is, conduct EFAs with and without outlier data to verify that the results are robust (Bandalos & Finney, 2019; Leys et al., 2018; Tabachnick & Fidell, 2019; Thompson, 2004).

Missing Data

Ideally, there will be no missing data. In practice, there often are: people sometimes skip items on tests or surveys, are absent on exam day, etc. First described by Rubin (1976), it is now accepted that the treatment of missing data is contingent on the mechanism that caused the data to be missing. Data that is missing completely at random (MCAR) is entirely unsystematic and not related to any other value on any measured variable. For example, a person might accidentally skip one question on a test. Data missing at random (MAR), contrary to its label, is not missing at random. Rather, it is a situation where the missingness can be fully accounted for by the remainder of the data. For instance, nonresponse to self-esteem questions might be related to other questions, such as gender and age. Finally, missing not at random (MNAR) applies when the missing data is related to the reason that it is missing. For example, an anxious person might not respond to survey questions dealing with anxiety.

It is useful to look for patterns at both variable and participant levels when considering missing data (Fernstad, 2019). For example, the first

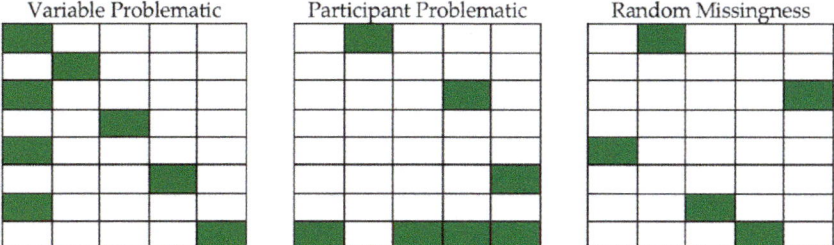

Figure 8.14 Missing data patterns

variable in the left panel in Figure 8.14 seems to be problematic whereas the final participant is problematic in the middle panel. The third panel depicts relatively random missingness. Generally, randomly scattered missing data is less problematic than other patterns (Tabachnick & Fidell, 2019).

Researchers tend to rely on two general approaches to dealing with missing data: discard a portion of the data or replace missing values with estimated or imputed values. When discarding data, the entire case can be discarded if one or more of its data values are missing (listwise deletion). Alternatively, only the actual missing values can be discarded (pairwise deletion). Most statistical programs offer these two missing data methods. Both methods will reduce power and may result in statistical estimation problems and biased parameter estimates (Zygmont & Smith, 2014). However, pairwise deletion is especially prone to statistical problems that create inadmissible EFA results (Lorenzo-Seva & Ferrando, 2021; Wothke, 1993).

A wide variety of methods have been developed to estimate or impute missing data values (Hair et al., 2019; Roth, 1994; Tabachnick & Fidell, 2019) that range from simple (replace missing values with the mean value of that variable) to more complex (predict the missing data value using nonmissing values via regression analysis) to extremely complex (multiple imputation and maximum likelihood estimation). Baraldi and Enders (2013) suggested that "researchers must formulate logical arguments that support a particular missing data mechanism and choose an analysis method that is most defensible, given their assumptions about missingness" (p. 639).

Unfortunately, there is no infallible way to statistically verify the missing data mechanism and most methods used to deal with missing data values rely, at a minimum, on the assumption of MAR. Considerable simulation research has suggested that the *amount* of missing data may be a practical guide to dealing with missing data. In general, if less than 5% to 10% of the data are missing in a random pattern across variables and participants, then any method of deletion or imputation will be acceptable (Chen et al., 2012; Hair et al., 2019; Lee & Ashton, 2007; Roth, 1994; Tabachnick & Fidell, 2019; Xiao et al., 2019). When more than 10% of the data are missing, Newman (2014) suggested that complex multivariate techniques, such as multiple imputation or maximum likelihood estimation, be used.

As with outliers, extensive missing data requires a sensitivity analysis where the EFA results from different methods of dealing with missing data are compared for robustness (Goldberg & Velicer, 2006; Hair et al., 2019; Tabachnick & Fidell, 2019). Large amounts of missing data might be the result of deficiencies in data collection procedures that should be addressed before further analyses are attempted (Goldberg & Velicer, 2006). Additionally, the amount and location of missing data at variable and participant levels should be transparently reported.

Missing data in Stata. Often, missing data are recognized during manual data entry and indicator values are deliberately assigned. For example, −9 assigned to missing values without apparent cause, −99 to missing values where the survey respondent refused to answer, and −999 when the question did not apply. Stata can recognize 27 numeric missing data indicators that will appear in the Data Editor from ".a" to ".z" and ".". Missing data indicators can be assigned with commands in several ways. Perhaps the most direct is variable by variable: `replace vocab1 = .a if vocab1 == -9`, `replace vocab1 = .b if vocab1 == -99`, and replace `replace vocab1 = .c if vocab1 == -999`. It would be easier to assign all three missing data indicators to all variables at once via `mvencode _all mv(-9 = .a \ -99 = .b \ -999 = .c)`. If there is only one missing data indicator, Stata will display "." for that value in the **Data Editor** window rather than the actual missing data indicator. Thus, "." is Stata's system missing data indicator.

The iq data set does not contain any missing data, but a version of that data set was created with ten random missing values (all indicated with −999) and imported via the menu sequence of ***File > Import > Excel spreadsheet > iqmiss.xlsx*** to demonstrate missing data in Stata. Reviewing the **Properties** pane, it appears that Stata saw the minus signs in the datafile and mistakenly specified these variables to be string types. That is, the data were seen as words instead of numbers. Menu options to convert the string type variables to numeric type variables are ***Data > Create or change data > Other variable-transformation commands > Convert variables from string to numeric > Convert specified variables to numeric***. A simple command accomplishes the same function: `destring vocab1-designs2, replace`.

Sometimes missing value indicators were assigned but Stata was not informed that those values are not real, but only indicators of missingness. It is important that data be screened to verify that missing values are appropriately indicated and handled. Type `browse` in the **Command** pane to review this data. Notice that −999 values are displayed, indicating that missing

variable	N	mean	sd	min	max	skewness	kurtosis
vocab1	152	83.19079	126.5562	-999	137	-8.301962	71.28792
designs1	152	83.29605	126.1936	-999	130	-8.377658	72.13791
similar1	152	103.5921	17.25552	55	145	-.4533611	3.307541
matrix1	152	84.92763	126.6478	-999	134	-8.324863	71.53885
veranal2	152	94.17105	90.46098	-999	134	-11.72141	142.169
vocab2	152	100.6316	16.41623	37	144	-.8117374	4.787527
matrix2	152	87.10526	126.856	-999	137	-8.334263	71.64322
designs2	152	93.36842	90.26816	-999	137	-11.77036	142.9673

Figure 8.15 Descriptive statistics with missing data indicators mistakenly considered to be real data

data indicators were not specified for this data. The `nmissing` command will display the number of missing values. In this case, it returns a "." to indicate that Stata has not categorized any values as missing. Consequently, Stata will consider −999 to be valid iq values and use them in subsequent computations. For example, compare the results in Figure 8.15 with those in Figure 8.4.

As demonstrated in Figure 8.15, values of −999 have been assumed to be real and used in computation of descriptive statistics, which are incorrect. Similar erroneous EFA results could be obtained if missing data indicators are improperly included in the analysis.

An appropriate missing data indicator must be entered for each variable to ensure that this does not occur. Using menu options, this can be accomplished by **Data > Create or change data > Other variable-transformation commands > Change numeric values to missing > vocab1–designs2 > −999.** The same task can be accomplished with the `mvdecode vocab1-designs2, mv(-999)` command.

A review of the **Data Editor** window demonstrates that the missing data are now indicated by "." When the descriptive statistics are displayed with the `sum` command, the resulting output now correctly reports that some cases contain missing data indicators (i.e., there are 150 cases for some variables, 152 for other variables, and 142 cases without any missing data) and the missing data indicators have not been used in computations as if they were real data. A table of missing values can be produced by the `misstable summarize` command and a plot of missing values can be generated by the `misstable pattern` command, which are displayed in Figure 8.16.

As displayed in the `misstable summarize` portion of Figure 8.16, six variables are each missing one or two data points across ten participants. In the `misstable patterns` portion of Figure 8.16, each row represents a set of observations that have the same pattern of missing data. For example, the first row represents the most common data pattern where there was no missing data (93%). The second row is a pattern where the

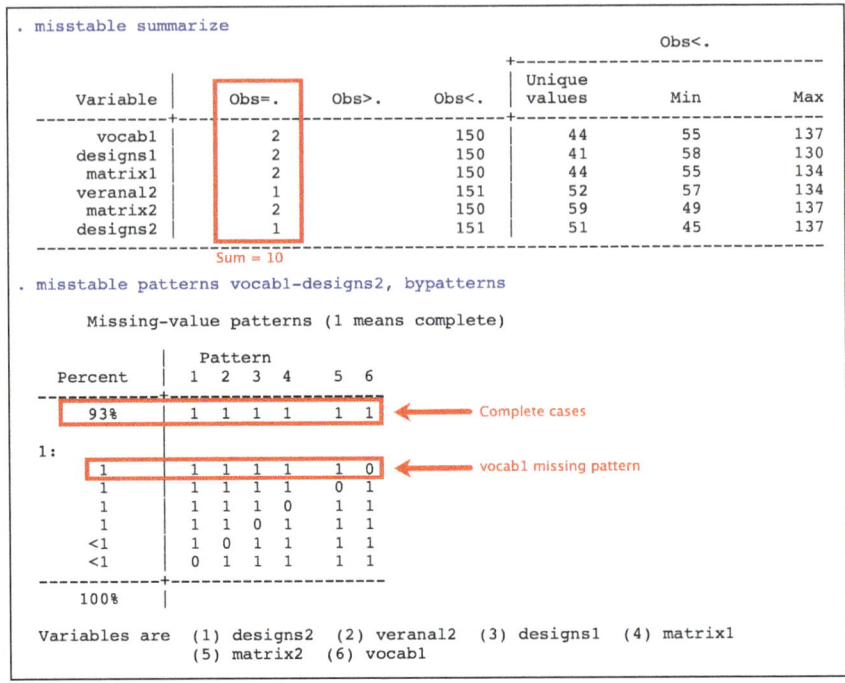

Figure 8.16 Results from misstable command

vocab1 variable was missing data but the other variables were complete and so on. With 152 cases and eight variables, there are 1,216 data points. Of those, 1,206 were complete and ten missing (0.8%). This is a trivial amount of missing data.

Other missing data enumerations can be obtained from user-supplied routines that can be downloaded. For example, ssc install mdesc, ssc install nmissing, and ssc install missingplot.

Data imputation in Stata. Conducting a factor analysis on this dataset without imputing missing data will result in a sample size of 142 because the underlying correlation routine in Stata uses listwise deletion. Although not recommended, mean substitution could be accomplished manually in the **Data Editor** window by typing in the mean values displayed in Figure 8.4.

A more complex predictive mean matching imputation can be performed with the user-contributed ice package (ssc install ice). In essence, this procedure uses regression methods to estimate the missing value of each variable and that variable for all other cases. It then imputes the observed value of the closest neighbor with a similar predicted value. Thus, this is

	vocab1	designs1	similar1	matrix1	veranal2	vocab2	matrix2	designs2
vocab1	1.0000							
designs1	0.5759	1.0000						
similar1	0.7886	0.5726	1.0000					
matrix1	0.6058	0.6668	0.5854	1.0000				
veranal2	0.6915	0.5108	0.7026	0.5308	1.0000			
vocab2	0.8241	0.5386	0.7353	0.5600	0.7075	1.0000		
matrix2	0.5662	0.5828	0.5759	0.7173	0.6490	0.5827	1.0000	
designs2	0.5097	0.6525	0.5521	0.6390	0.5187	0.5308	0.6275	1.0000

Figure 8.17 Descriptive statistics output of iqmiss data after imputation

a mixture of regression and hot-deck methods (Roth, 1994). Stata's `mi impute chained` command provides another method for imputation of missing data.

This procedure can be implemented with the `ice vocab1-designs2, match clear` command. Open the Data Editor window and observe that two new variables have been added (_mi and _mj) and that there are now 304 cases. The _mi variable is the replicated case number and the _mj variable identifies the imputation sequence with zero being the original data. The first 152 cases are the original cases with missing data and the second 152 are the imputed data. Simply delete cases 1–152 with the `drop in 1/152` command or the **Data > Create or change data > Drop or keep observations > Drop observations > Drop a range of observations from 1 to 152** menu options and use that imputed data for the subsequent EFA.

Compare the correlation matrices of this imputed data set (Figure 8.17) and the full dataset (Figure 8.5). The differences are minor and will not affect EFA results.

Currently, maximum likelihood expectation–maximization (EM), full information maximum likelihood (FIML), and multiple imputation (MI) are the most appropriate methods to apply when there is more than 5% to 10% missing data (Enders, 2017). The MI method can be implemented in Stata but does not synchronize with EFA commands unless a complex command using the `cmdok` option is employed (`help cmdok` for instructions). Likewise, the FIML method is inconvenient because it requires use of Stata's structural equation modeling routines. The EM method is based on a model described by Weaver and Maxwell (2014) and an example provided on the UCLA Statistical Consulting website. The do-file for that method is provided in Figure 8.18.

Any type of imputation should be acceptable, given the trivial amount of missing data involved in this analysis (Chen et al., 2012; Hair et al., 2019; Lee & Ashton, 2007; Roth, 1994; Tabachnick & Fidell, 2019; Xiao et al., 2019).

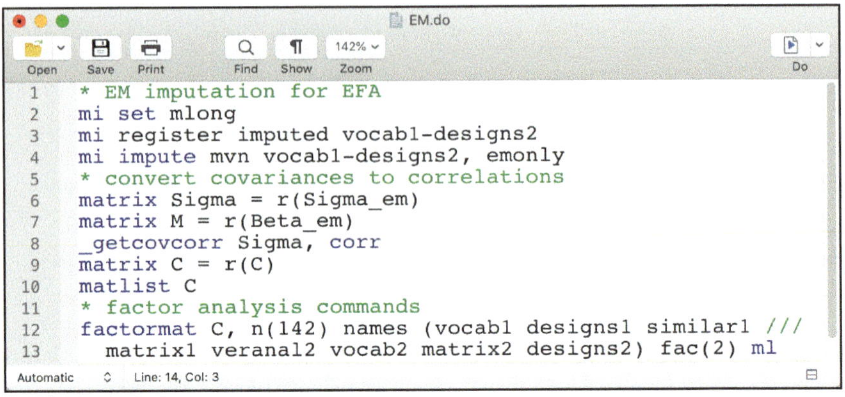

```
                                      EM.do
  Open   Save   Print        Find  Show  Zoom                        Do
  1   * EM imputation for EFA
  2   mi set mlong
  3   mi register imputed vocab1-designs2
  4   mi impute mvn vocab1-designs2, emonly
  5   * convert covariances to correlations
  6   matrix Sigma = r(Sigma_em)
  7   matrix M = r(Beta_em)
  8   _getcovcorr Sigma, corr
  9   matrix C = r(C)
  10  matlist C
  11  * factor analysis commands
  12  factormat C, n(142) names (vocab1 designs1 similar1 ///
  13     matrix1 veranal2 vocab2 matrix2 designs2) fac(2) ml
  Automatic   ○   Line: 14, Col: 3
```

Figure 8.18 Do-file for maximum likelihood expectation–maximization imputation

Report

Scatterplots revealed that linear relationships exist between the variables. Measures of univariate and multivariate normality indicated a relatively normal data distribution (Curran et al., 1996; Finney & DiStefano, 2013; Mardia, 1970). There was no evidence that restriction of range or outliers substantially affected the scores and there was no missing data. Therefore, a Pearson product–moment correlation matrix was submitted for EFA.

9 Step 4

Is Exploratory Factor Analysis Appropriate?

Given that exploratory factor analysis (EFA) is based on the correlation matrix, it seems reasonable that the correlation matrix should contain enough covariance to justify conducting an EFA (Dziuban & Shirkey, 1974). First, the correlation matrix can be visually scanned to ensure that there are several coefficients $\geq$ .30 (Hair et al., 2019; Sheskin, 2011; Tabachnick & Fidell, 2019; Yong & Pearce, 2013).

As was illustrated in Figure 8.5, it took only a single command to produce a correlation matrix for the small iq dataset. A quick visual scan of this 8 by 8 matrix reveals many coefficients $\geq$ .30. The smallest coefficient is .5074 and the largest is .8242.

Scanning a large matrix for coefficients $\geq$ .30 is laborious but can be automated with a user-contributed ado-file called corrci that can be located with the `search corrci` command. The resulting **Viewer** window displays three links in reverse chronological order. The first listing is the most recent update of the corrci ado-file. Click that link and the *click here to install* link on the next page. Figure 9.1 displays a do-file that will compute correlations and their confidence intervals as well as the minimum, maximum, and average correlation coefficient using the `corrci` command.

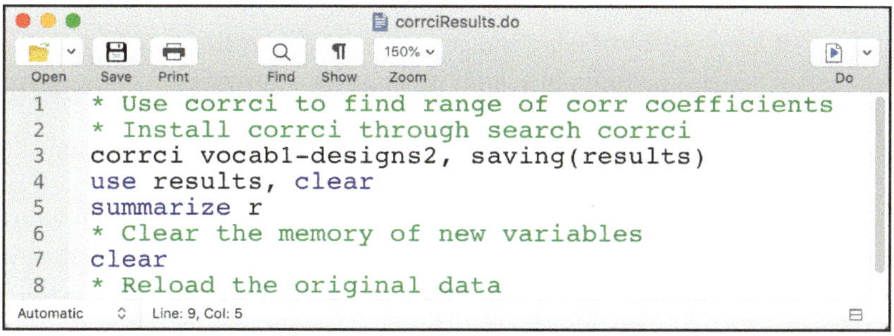

```
     corrciResults.do
     Open  Save  Print      Find  Show  150% ∨        Zoom                    Do
1    * Use corrci to find range of corr coefficients
2    * Install corrci through search corrci
3    corrci vocab1-designs2, saving(results)
4    use results, clear
5    summarize r
6    * Clear the memory of new variables
7    clear
8    * Reload the original data
Automatic    ◇   Line: 9, Col: 5
```

Figure 9.1 Stata corrci do-file

DOI: 10.4324/9781003149286-9

```
                      correlations and 95% limits
vocab1    designs1        0.576       0.459      0.674
vocab1    similar1        0.789       0.720      0.842
vocab1    matrix1         0.618       0.509      0.707
vocab1    verana12        0.689       0.595      0.765
vocab1    vocab2          0.824       0.765      0.869
vocab1    matrix2         0.564       0.444      0.663
vocab1    designs2        0.507       0.379      0.617
designs1  similar1        0.570       0.451      0.668
designs1  matrix1         0.650       0.547      0.733
designs1  verana12        0.509       0.380      0.618
designs1  vocab2          0.539       0.415      0.643
designs1  matrix2         0.586       0.471      0.682
designs1  designs2        0.662       0.561      0.743
similar1  matrix1         0.598       0.486      0.692
similar1  verana12        0.702       0.611      0.774
similar1  vocab2          0.735       0.652      0.801
similar1  matrix2         0.582       0.466      0.678
similar1  designs2        0.550       0.429      0.652
matrix1   verana12        0.533       0.408      0.638
matrix1   vocab2          0.566       0.447      0.665
matrix1   matrix2         0.715       0.627      0.785
matrix1   designs2        0.619       0.510      0.708
verana12  vocab2          0.706       0.616      0.778
verana12  matrix2         0.648       0.545      0.732
verana12  designs2        0.514       0.386      0.622
vocab2    matrix2         0.577       0.460      0.675
vocab2    designs2        0.530       0.405      0.636
matrix2   designs2        0.624       0.517      0.713

    Variable |     Obs      Mean    Std. Dev.       Min         Max
-------------+-------------------------------------------------------------
           r |      28   .6172034   .0845934    .5073953    .8241553
```

Figure 9.2 Results of corrci do-file

The `corrci` command computes the correlation coefficient and its confidence interval, creates a set of new variables, clears the iq variables from memory, and saves a summary in a file named "results" in the working directory. The subsequent `summarize r` command displays the minimum, maximum, and average correlation coefficient (Figure 9.2).

The appropriateness of a correlation matrix for EFA can be evaluated with a community-contributed ado-file named factortest that can be installed via `ssc install factortest` and implemented with the `factortest vocab1-designs2` command. The results of that command are displayed in Figure 9.3.

The likelihood of a multicollinearity problem can be checked by ascertaining the determinant of the correlation matrix. The determinant of a singular matrix (i.e., one that cannot be inverted) is zero and the determinant of a matrix of orthogonal variables is one (Pett et al., 2003; Rockwell, 1975). Thus, a determinant somewhere between zero and one is required for the mathematical operations of EFA. Field et al. (2012) suggested that if the determinant is greater than .00001 then multicollinearity is probably not a problem. Using that standard, a value of .002 for this data indicates that multicollinearity is not a serious issue.

```
. factortest vocabl-designs2

Determinant of the correlation matrix
Det                =       0.002


Bartlett test of sphericity

Chi-square         =              887.074
Degrees of freedom =                   28
p-value            =                0.000
HO: variables are not intercorrelated


Kaiser-Meyer-Olkin Measure of Sampling Adequacy
KMO                =       0.903
```

Figure 9.3 Determinant, Bartlett test of sphericity, and KMO for iq data via factortest command

Haitovsky (1969) suggested a null hypothesis test to determine if a matrix is singular. Although not provided by Stata, a standalone computer program to compute Haitovsky's statistical test is available at edpsychassociates.com. As revealed in Figure 9.4, it is highly unlikely that this matrix is singular so it should be appropriate for EFA.

A second assessment of the appropriateness of the correlation matrix for EFA is provided by Bartlett's (1950) test of sphericity, which tests the null hypothesis that the correlation matrix is an identify matrix (i.e., random with ones on the diagonal and zeros on the off-diagonal) in the population similar to the Haitovsky test. The desired outcome is rejection of the random matrix with a statistically significant chi-square test. Bartlett's test is sensitive to sample size and should be considered a minimal standard (Nunnally & Bernstein, 1994). The chi-square value for this correlation matrix was 887.1 with 28 degrees of freedom, indicating that it is not a random matrix. Stata reported a value of .000 but that does not mean that *p* is zero, only that it is significant to three decimal places.

A final assessment of the appropriateness of the correlation matrix for EFA is offered by the Kaiser–Meyer–Olkin measure of sampling adequacy (KMO; Kaiser, 1974). KMO values range from 0 to 1, only reaching 1 when each variable is perfectly predicted by the other variables. A KMO value is a ratio of the sum of squared correlations to the sum of squared correlations plus the sum of squared partial correlations. Essentially, partial correlations will be small and KMO values large if the items share common variance. Kaiser (1974) suggested that KMO values < .50 are unacceptable but other measurement specialists recommended a minimum value of .60 (Mvududu & Sink, 2013; Sheskin, 2011; Watson, 2017) for acceptability with values ≥

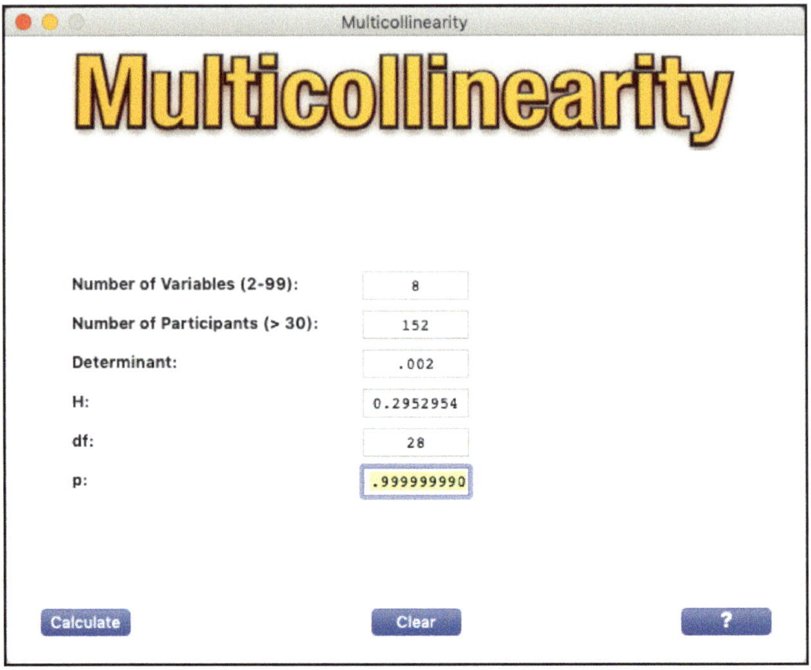

Figure 9.4 Multicollinearity program to compute Haitovsky's singularity test

.70 preferred (Hoelzle & Meyer, 2013). KMO values for each variable as well as the overall average should be reviewed. If no KMO values can be computed or if they fall outside the theoretical range of 0 to 1, the correlation matrix may not be positive definite and should be reviewed for accuracy and appropriateness (Lorenzo-Seva & Ferrando, 2021). In this case, the KMO value of .903 is high and acceptable.

The KMO value of .903 is an average for all the variables. It is possible that the overall average is acceptable although the sampling adequacy of one or more variables might be unacceptable. The KMO value for each variable can be produced with another command, producing the KMO values for each variable. However, Stata only produces that matrix after an EFA has been conducted. When implemented following an EFA, the `estat kmo` command produced KMO values of .866 to .930, indicating that the overall KMO value of .903 is a fair representation of the data.

Report

A visual scan of the correlation matrix for the data revealed that all of the coefficients were ≥ .30, but none exceeded .90 (Tabachnick & Fidell, 2019). Bartlett's test of sphericity (1950) rejected the hypothesis that the correlation matrix was an identity matrix (chi-square of 887.1 with 28 degrees of freedom). The Kaiser–Meyer–Olkin (KMO) measure of sampling adequacy was acceptable with values of .90 for the total model and .87 to .93 for each of the measured variables (Kaiser, 1974). Altogether, these measures indicate that the correlation matrix is appropriate for EFA (Hair et al., 2019; Tabachnick & Fidell, 2019).

10 Step 5

Factor Analysis Model

Two major models must be considered: principal components analysis (PCA) and common factor analysis. Only common factor analysis can be classified as an exploratory factor analysis (EFA) technique. Researchers sometimes claim that an EFA was conducted when a PCA model was actually applied (Osborne & Banjanovic, 2016). However, PCA and EFA have different purposes and might, given the number and type of measured variables, produce different results.

EFA is based on the common factor model described in the Introduction (Fabrigar & Wegener, 2012) and is pictured in Figure 10.1. The purpose of EFA is to explain as well as possible the correlations, or covariance, among measured variables (Sheskin, 2011). In EFA, measured variables are thought to correlate with each other due to underlying latent constructs called factors. The direction of influence from factor to measured variables is signified by the arrows in the path diagram. This model also assumes that unique factors explain some variance beyond that explained by common factors. Conceptually, unique factors are composed of specific variance (systematic

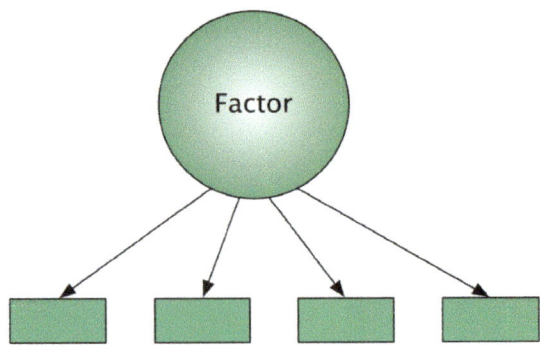

Figure 10.1 Common factor analysis model

DOI: 10.4324/9781003149286-10

Principal Components Analysis

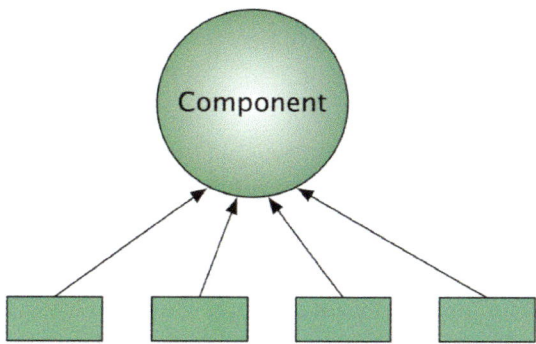

Observed Variable · Measured Variable · Manifest Variable · Formative Indicator

Figure 10.2 Principal components analysis model

variance specific to a single measured variable) and error variance (unreliable error of measurement).

The purpose of PCA is to take the scores on a large set of observed variables and reduce them to scores on a smaller set of composite variables that retain as much information as possible from the original measured variables (Sheskin, 2011). The direction of influence in PCA is from measured variables to factors, which is symbolized by the arrows in the path diagram in Figure 10.2. PCA attempts to explain as much variance as possible and does not differentiate between common (shared) variance and unique (systematic and error) variance. PCA evaluates variance, not covariance. Therefore, principal components are *not* latent variables. Linear functions are properly called components, not factors.

As summarized in Figure 10.3, the common factor model (called EFA for our purposes) is composed of common variance (general variance shared by all measured variables plus variance shared by subsets of the measured variables) plus unique variance (variance specific to a single measured variable plus error). In contrast, the PCA model does not distinguish common

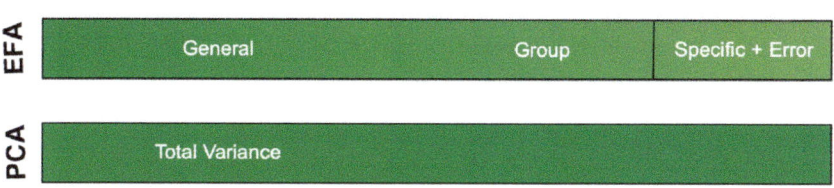

Figure 10.3 Variance components of EFA and PCA models

variance from unique variance, analyzing the total variance of the measured variables similar to multiple regression.

The relative advantages and disadvantages of EFA and PCA models have long been debated. Widaman (1993) found more bias in PCA estimates than in EFA estimates and Gorsuch (1990) recommended that EFA "should be routinely applied as the standard analysis because it recognizes we have error in our variables, gives unbiased instead of inflated loadings, and is more elegant as a part of the standard model" (p. 39). Other methodologists have argued that EFA and PCA tend to produce similar results so it does not matter which is used (Hoelzle & Meyer, 2013; Velicer & Jackson, 1990).

In truth, PCA and EFA may sometimes produce similar results. However, that is dependent on the number of variables involved in the analysis and the amount of common variance shared by the measured variables. The computation of PCA and EFA differs in how the diagonal of the correlation matrix is handled. In PCA, the correlations of 1.00 between each variable and itself are used. In contrast, EFA begins by replacing the 1s on the diagonal with an estimate of the communality or common variance. This is called a reduced correlation matrix. Thus, PCA considers all the variance (common and error) whereas EFA considers only the common variance. Because it is only the diagonal elements of the correlation matrix that differ, the number of diagonal elements influences the difference between EFA and PCA results. For example, there are eight diagonal elements and 28 nonredundant off-diagonal elements for eight measured variables for a 22% difference in the models, but 20 diagonal elements and 190 nonredundant off-diagonal elements for 20 measured variables is a difference of 10% between EFA and PCA. Thus, EFA and PCA results will tend to be more similar when there are more measured variables in the analysis (10% versus 22% different correlation elements).

As noted, the common factor model recognizes the presence of error in all measurements and therefore substitutes estimates of variable communalities (instead of 1s) in the diagonal of the correlation matrix. Unfortunately, it is not possible to know the communality of variables before conducting an EFA. The solution to this unknown is to estimate the communalities based on some aspect of the data. Potential solutions include the reliability of the variables, partial correlations, multiple correlations, etc. Over time, it has become accepted that a good solution is to initially estimate communalities with the squared multiple correlation (SMC) of each variable with all other variables and then systematically refine that estimate through a series of iterations until a stable estimate is reached (Fabrigar & Wegener, 2012; Pett et al., 2003; Tabachnick & Fidell, 2019).

Widaman (2018) concluded that "PCA should never be used if the goal is to understand and represent the latent structure of a domain; only FA techniques should be used for this purpose" (p. 829). Similar opinions were expressed by Bandalos (2018); Bandalos and Boehm-Kaufman (2009); Carroll (1978, 1985); Fabrigar et al. (1999); Fabrigar & Wegener, 2012; Finch

(2013); Haig (2018); Hair et al. (2019); Preacher and MacCallum (2003); Russell (2002); and Schmitt (2011).

Report

The purpose of this study was to uncover the latent structure underlying these eight measured variables. Accordingly, a common factor model (EFA) was selected (Widaman, 2018). Squared multiple correlations (SMC) were used for initial communality estimates (Tabachnick & Fidell, 2019).

11 Step 6

Factor Extraction Method

After selecting the common factor model, the next step in exploratory factor analysis (EFA) is to choose an extraction method. This might also be called the model fitting procedure or the parameter estimation procedure (Fabrigar & Wegener, 2012). In simple terms, this is the mathematical process of deriving the underlying factors from the correlation matrix.

In current practice, the EFA software applies mathematical routines to complete factor extraction. Before computers, extraction was completed by humans who used a geometric approach. That approach retains conceptual clarity and a simple case of two variables will be used for pedagogic purposes.

Data Space

Beginning with a correlation matrix, a scatterplot can be used to display the relationship between X and Y variables in two-dimensional space as in Figure 11.1. Each data point represents an individual's unit standardized scores on variables X and Y (i.e., z scores). The center is now the mean of both variables. Notice that X–Y score pairs fall into four quadrants. Score pairs which fall into quadrants 1 and 3 tend toward positive correlations, those in quadrants 2 and 4 tend toward negative, and if evenly distributed across all four quadrants then correlation will be zero. This is a geometric display of what could be called *data space*. More variables would necessitate a view of multidimensional space that is difficult to visualize.

Factor Space

One geometric way to identify the factor(s) underlying a correlation matrix is to convert from data space to *factor space*. The first step in conversion to factor space is to identify the one straight line that lies closer to all the data points than any other line. This is called the "least squares best-fitting line" because the sum of the squared deviations of each of the data points from one and only one straight line is at a minimum. This particular best-fitting line is the first principal component if extracted from the full correlation

DOI: 10.4324/9781003149286-11

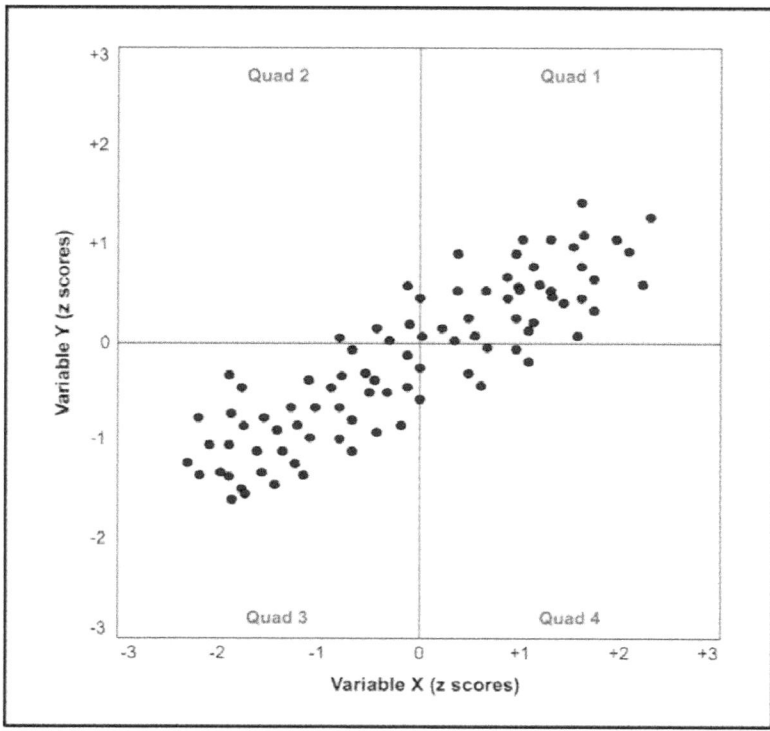

Figure 11.1 Scatterplot of two standardized variables in data space

matrix and the first principal factor if extracted from the reduced correlation matrix. This first component or factor "accounts for" more of the variance or covariance than any other possible component or factor. This best-fitting line is designated by the Roman numeral I. By convention, these lines are scaled from −1.00 to +1.00.

Each individual X–Y data point can be converted from its X–Y location in data space to a location in factor space by connecting the data point to line I with an orthogonal projection. As depicted in Figure 11.2, the new location is approximately 0.77 on Factor I.

The second principal component or factor, labeled II, is defined as a straight line at a right angle to the first principal component/factor. It too is a best-fitting straight line, and it accounts for that part of the total variance of the test score that was *not* accounted for by the first principal component/factor. Being at right angles, the two lines are uncorrelated. The X–Y data point is converted to a location on Factor II by connecting the data point to line II with an orthogonal projection. The factor II location appears to be near −0.43. If this analysis had included more measured variables, additional

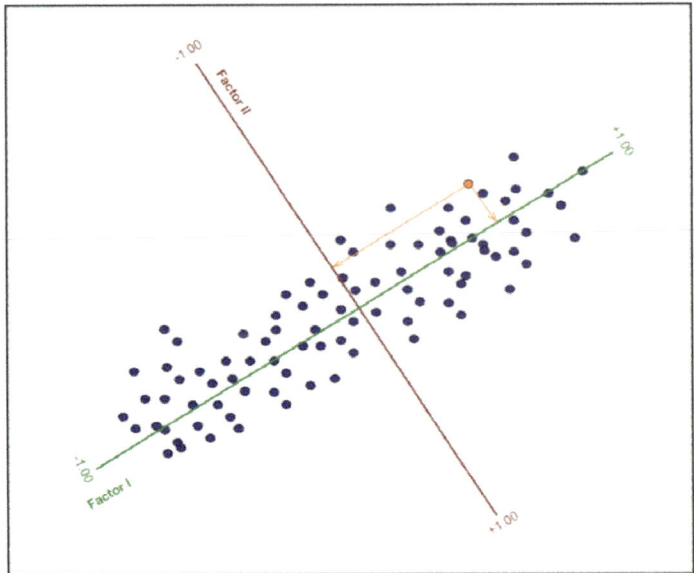

Figure 11.2 Conversion from data space to factor space

factors would have been extracted with each succeeding factor accounting for a smaller proportion of variance analogous to wringing water from a wet towel.

In this simplest conceptual example of the principal components or principal factor (also called principal axis) variant of factor analysis, one set of references (X and Y) was exchanged for another set of references (I and II). The same amount of variance (in this case, the total or reduced variance of X and Y) is accounted for by both sets of references. The only difference is that X and Y are correlated variables, whereas I and II are uncorrelated components or factors. For now, the factor space references will be called factor loadings.

Eigenvalues, Variance, and Communality

Theoretically, as many factors as measured variables can be extracted. Each variable can be located in factor space with its unique factor loadings. Typically, this information is presented in a table where the factors, conventionally identified with Roman numerals, and measured variables are entered in columns and rows, respectively, as in Figure 11.3. In practice, parsimony often suggests that most of the variance can be accounted for by only a few factors; for example, where two factors are sufficient to account for the majority of variance in five measured variables.

| | Unrotated Factors | | |
	I	II	h^2
Variables			
A	.73	.54	.825
B	.82	.49	.913
C	.72	.69	.995
D	.77	-.43	.778
E	.84	-.44	.899
Eigenvalue	3.02	1.39	4.41
% Total variance	60.40	27.80	88.20
Cumm % total var	60.40	88.20	
% Common variance	68.48	31.52	

Figure 11.3 Sample EFA table

The proportion of each variable's variance accounted for by a factor is the square of that variable's loading on the factor (e.g., Variable A loading of $.73^2 = .533$ for Factor I and $.54^2 = .292$ for Factor II). Communality is symbolized by h^2. It is the proportion of variance in each variable that is accounted for by all factors (e.g., $.533 + .292 = .825$). The uniqueness of a variable is defined as one minus the communality of the variables (e.g., $1 - .825 = .175$). Uniqueness is comprised of reliable variance specific to a variable and unreliable error variance. Theoretically, the proportion of error variance can be identified by subtracting the reliability of the variable from one. Given these definitions, high communalities (low uniqueness) are desired and communalities $\geq .60$ are often considered to be high (Gibson et al., 2020).

An eigenvalue is an algebraic solution for the percentage of variance a factor contributes and can be thought of as the equivalent number of variables a factor represents. Sometimes called the latent root or characteristic root, the eigenvalue is the sum of the factor's squared loadings (e.g., $.73^2 + .82^2 + .72^2 + .77^2 + .84^2 = 3.02$). Thus, this first factor accounts for as much variance as three variables. An eigenvalue is one measure of the relative importance of each factor (Sheskin, 2011).

Mathematically, the total variance of a set of variables is equal to the total number of variables. Thus, the total variance for this five variable EFA is 5.00. Based on this sum, the proportion of the total variance contributed by a factor is its eigenvalue divided by the total variance (e.g., $3.02 \div 5.00 = .604$ or 60.40%). The proportion of common variance contributed by a factor is its eigenvalue divided by the sum of eigenvalues (e.g., $3.02 \div 4.41 = .6848$ or

68.48%).The sum of the communalities is equal to the sum of the eigenvalues (e.g., 3.02 + 1.39 = 4.41).

Extraction Methods

Methodologists have devised a great number of extraction methods since Spearman (1904) first imagined EFA. These include image analysis, alpha analysis, non–iterated principal axis (PA), iterated principal axis (IPA), maximum likelihood (ML), unweighted least squares (ULS), weighted least squares (WLS), generalized least squares (GLS), ordinary least squares (OLS), minimum residual (MINRES), etc. Some extraction methods are fundamentally identical yet were given different names. For example, ULS, OLS, and MINRES all refer to essentially the same method (Flora, 2018). Further, an IPA extraction will converge to an OLS/ULS solution (Briggs & MacCallum, 2003; MacCallum, 2009). Mathematically, extraction methods differ in the way they go about locating the factors that will best reproduce the original correlation matrix, whether they attempt to reproduce the sample correlation matrix or the population correlation matrix, and in their definition of the best way to measure closeness of the reproduced and original correlation matrices.

The most significant distinction between common factor extraction methods is between ML and least-squares methods (OLS/ULS, GLS, and IPA). ML attempts to reproduce the *population* correlation matrix that most likely generated the sample correlation matrix, whereas the least–squares methods attempt to reproduce the *sample* correlation matrix. The population focus of ML extraction leaves it dependent on two critical assumptions: (a) the data are from a random sample of some population, and (b) the measured variables have an underlying multivariate normal distribution. In contrast, the least-squares methods have no distributional assumptions (Fabrigar & Wegener, 2012). IPA seeks to maximize the common variance extracted by forming a weighted combination of all the variables that will produce the highest squared correlations between the variables and the factor, whereas OLS/ULS attempts to minimize the differences between the off-diagonal elements of the correlation matrix and the reproduced correlation matrix. IPA and OLS/ULS both produce "the best least-squares estimate of the entire correlation matrix including diagonal elements" (Gorsuch, 1983, p. 96). Generalized least squares (sometimes called weighted least squares) modifies the ULS method by giving more weight to variables with high correlations. OLS/ULS is probably most appropriate when non–positive definite matrices are encountered with another extraction method (Lorenzo-Seva & Ferrando, 2021).

Stata offers four extraction methods: principal factor (PF), principal component factor (PCF), iterated principal factor (IPF), and maximum likelihood (ML). Contrary to its label, the principal component factor option is just the classical principal components method that produces components, not factors. The principal factor and iterated principal factor methods are similar except the iterated principal factor method reestimates the communalities iteratively and the principal factor method does not. As a default, the IPA method will reestimate communality estimates until the change is .0001 or less, but users can specify the number of iterations allowed by including that number in the `ipf citerate(2)` command. Stata uses the SMC as the default communality estimate when the principal factor methods are selected.

Simulation research has compared ML and least-squares extraction methods in terms of factor recovery under varying levels of sample size and factor strength (Briggs & MacCallum, 2003; de Winter & Dodou, 2012; MacCallum et al., 2007; Ximénez, 2009). In general, least-squares methods have outperformed ML when the factors were relatively weak (i.e., the factor accounted for ≤ 16% of the variance of a measured variable), the model was wrong (too many factors were specified), and the sample size was small ($N = 100$). Given these results, Briggs and MacCallum (2003) recommended "use of OLS in exploratory factor analysis in practice to increase the likelihood that all major common factors are recovered" (p. 54). Thus, ML may be appropriate for larger samples with normal data (i.e., univariate skew ≤ 2.0 and kurtosis ≤ 7.0; multivariate kurtosis nonsignificant and ≤ 5.0) and strong factors, whereas least-squares methods may be preferable for smaller samples with nonnormal data or weak factors (MacCallum et al., 2007; Watson, 2017).

Some researchers prefer ML extraction because it attempts to generalize to the population and it allows computation of statistical tests of model parameters (Fabrigar et al., 1999; Fabrigar & Wegener, 2012; Matsunaga, 2010). Other researchers prefer least-squares extraction methods because they have no distributional assumptions and are sensitive to weak factors (Carroll, 1985, 1993; McCoach et al., 2013; Pett et al., 2003; Russell, 2002; Widaman, 2012). Osborne and Banjanovic (2016) concluded that "there is a general consensus in the literature that ML is the preferred choice for when data exhibits multivariate normality and iterated PAF or ULS for when that assumption is violated" (p. 26). Other researchers have endorsed that conclusion (Bandalos & Gerstner, 2016; Costello & Osborne, 2005; Sakaluk & Short, 2017; Schmitt, 2011; Yong & Pearce, 2013). However, different extraction methods tend to produce similar results in most cases (Tabachnick & Fidell, 2019).

Regardless of extraction method, researchers must be cautious of improper solutions (Lorenzo-Seva & Ferrando, 2021; van Driel, 1978; Wothke, 1993). Researchers must also be aware that extraction may fail with iterative estimators such as IPA. This happens because EFA software will try to arrive at an optimal estimate before some maximum number of estimation iterations has been completed. If an optimal estimate has not been computed at that point, an error message about nonconvergence will be produced. In that case, the researcher may increase the maximum number of iterations and rerun the analysis. If nonconvergence persists after several hundred iterations, then the results, like those from an improper solution, should not be interpreted (Flora, 2018). "Improper solutions and nonconvergence are more likely to occur when there is a linear dependence among observed variables, when the model includes too many common factors, or when the sample size is too small" (Flora, 2018, p. 257). Alternatively, another extraction method could be employed (e.g., IPA instead of ML).

Report

The univariate normal distribution of the data and substantial reliability of the measured variables indicated that a maximum likelihood extraction method might be appropriate. Nevertheless, the robustness of ML results will be verified by applying an iterated principal factor method to ensure that weak factors are not overlooked (Briggs & MacCallum, 2003; de Winter & Dodou, 2012).

12 Step 7

How Many Factors to Retain

As many factors as measured variables can be extracted in an exploratory factor analysis (EFA) but it is usually possible to explain the majority of covariance with a smaller number of factors. The problem arises in determining the *exact* number of factors to retain for interpretation. Methodologists have observed that this is probably the most important decision in EFA because there are serious consequences for selecting either too few or too many factors (Benson & Nasser, 1998; Fabrigar et al., 1999; Glorfeld, 1995; Hoelzle & Meyer, 2013; Preacher et al., 2013), whereas the options for other EFA decisions tend to be fairly robust (Hayton et al., 2004; Tabachnick & Fidell, 2019). Retaining too few factors can distort factor loadings and result in solutions in which common factors are combined, thereby obscuring the true factor solution. Extracting too many factors can focus on small, unimportant factors that are difficult to interpret and unlikely to replicate (Hayton et al., 2004). "Choosing the number of factors is something like focusing a microscope. Too high or too low an adjustment will obscure a structure that is obvious when the adjustment is just right" (Hair et al., 2019, p. 144).

"In the end, the overriding criteria for choosing a particular number of factors are interpretability and theoretical relevance because a factor solution is useful only if it can be interpreted in a meaningful way" (Bandalos, 2018, p. 324). A delicate balance between *comprehensiveness* and *parsimony* is needed to achieve an interpretable solution. Unfortunately, no infallible method to determine the "true" number of factors to retain has been discovered (Bandalos, 2018; Barrett & Kline, 1982; Cattell, 1966; Comrey & Lee, 1992; Fabrigar & Wegener 2012; Gorsuch, 1983; Nunnally & Bernstein, 1994; Pett et al., 2003; Preacher et al., 2013; Rummel, 1970; Widaman, 2012). It appears that the nature of the data (e.g., the number of indicators per factor, communality level, factor intercorrelations, complex loadings, sample size, etc.) differentially affects each method. Given this uncertainty, methodologists have recommended that both empirical evidence and theoretical knowledge be applied to determine the number of factors to retain for interpretation (Bandalos, 2018; Finch, 2020a; Hair et al., 2019; Osborne, 2014; Pituch & Stevens, 2016). Additionally, it has generally been agreed that underextraction is more dangerous than overextraction, so it may not be a bad strategy to risk the overextraction of one or two factors (Cattell, 1978; Fava & Velicer, 1992,

DOI: 10.4324/9781003149286-12

1996; Gorsuch, 1983; Kline, 2013; MacCallum et al., 2001; Stewart, 1981; Wood et al., 1996).

Empirical Guidelines

A variety of simulation studies have provided empirical guidelines for determining the number of factors to interpret. Based on these simulations, some empirical methods have been found to perform better than others.

Parallel Analysis

The parallel analysis (PA) criterion of Horn (1965) has generally performed well (Auerswald & Moshagen, 2019; Finch, 2020b; Peres-Neto et al., 2005; Ruscio & Roche, 2012; Velicer et al., 2000; Zwick & Velicer, 1986). Conceptually, PA involves generating a set of random data with the same number of variables and participants as the real data and then comparing the mean eigenvalues from multiple sets of random data with the corresponding eigenvalues from the real data. In this comparison, only factors with eigenvalues that are above the mean of those from the random data should be retained.

PA has remained accurate with nonnormal data and non-Pearsonian correlations (Buja & Eyuboglu, 1992; Dinno, 2009; Garrido et al., 2013; Li et al., 2020) and has performed better than modified PA versions (Lim & Jahng, 2019). Some methodologists have noted that PA tends to overextract by one or two factors and recommended that the upper 95th or 99th percentile of each eigenvalue be retained rather than the mean (Glorfeld, 1995; Hoyle & Duvall, 2004). However, Crawford et al. (2010) found that PA tended to underextract if there was a strong general factor and Caron (2019) reported that PA also tended to underextract if there were correlated factors. Given these competing results, it seems reasonable to compare mean eigenvalues rather than risk underextraction. PA is easily computed with modern computers so it is recommended that at least 100 random datasets be generated (Hoelzle & Meyer, 2013).

Minimum Average Partial

A second empirical guideline that has been found to be relatively accurate in simulation studies is the minimum average partial (MAP) method of Velicer (1976). A matrix of partial correlations is calculated after each principal component is extracted. The average of the squared off-diagonal partial correlations is computed from each matrix. This average is expected to reach a minimum when the correct number of components is extracted. The logic of this procedure is that as common variance is partialled out of the matrix, the MAP criterion will decrease. At the point where the common variance has been removed and only unique variance remains, the MAP criterion will begin to rise. Thus, the MAP criterion separates common and unique vari ance and retains only components consisting of common variance.

Scree Plot

A third, more subjective, method relies on graphing the eigenvalues derived from extraction and using visual analysis to detect any abrupt change in their slope. Developed by Cattell (1966), this scree graph plots the eigenvalues on the Y (vertical) axis against their extraction order on the X (horizontal) axis. The concept is that the eigenvalues for the "true" common factors tend to form a more or less straight line, whereas the error factors form another line with a different, smaller slope (analogous to the scree or rubble at the base of a cliff). This assumes that as succeeding factors are extracted from the factor matrix, error variance will begin to predominate and will represent only minor, random fluctuations. To follow the geology analogy, this will separate important early factors (bedrock) from the rubble (scree) of random error. Some methodologists supplement the scree with a minimum variance guideline, but these guidelines are subjective (e.g., 40% to 75%) and unlikely to be reasonable for social science research (Beavers et al., 2012 Reio & Shuck, 2015).

An idealized scree plot is illustrated in Figure 12.1. To analyze a scree plot, draw a straight line through the eigenvalues from right to left. The eigenvalues above and to the left of the straight line are the factors to retain (Pett et al., 2003). This example is relatively easy to interpret. The three largest eigenvalues are clearly above the trajectory of the straight line. However, most scree plots are more ambiguous and researchers tend to be unreliable when using scree plots to determine the number of factors to retain (Streiner, 1998).

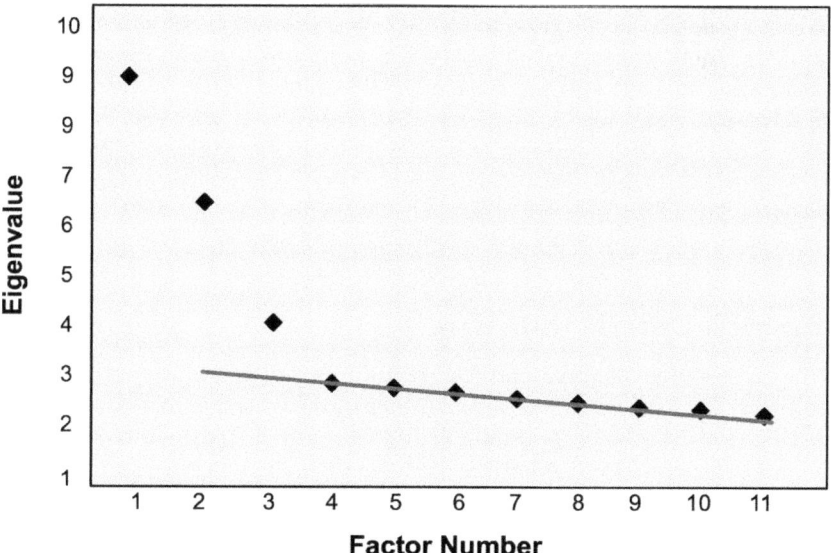

Figure 12.1 Simplified scree plot

Attempts have been made to interpret scree plots more objectively via empirical methods. These include the optimal coordinates method (Raîche et al., 2013) that uses eigenvalues as predictors in a multiple regression and identifies the cut point when an observed eigenvalue exceeds its estimated value. Another empirical approach is the acceleration factor (Yakovitz & Szidarovszky, 1986) that identifies where the slope in the curve of the scree plot changes abruptly. Finally, the standard error scree (Zoski & Jurs, 1996) is based on the standard error of the estimate of each eigenvalue. Unfortunately, results from simulation studies have been inconsistent and it is not clear if any empirical scree method is superior (Nasser et al., 2002; Raîche et al., 2013).

Other Empirical Guidelines

Numerous other empirical guidelines have been suggested by researchers. For example, innovative network graphing methods (Golino & Epskamp, 2017; Golino et al., 2020), a very simple structure (VSS) approach that minimizes factor complexity (Revelle & Rocklin, 1979), the Hull method that attempts to balance goodness-of-fit and parsimony (Lorenzo-Seva et al., 2011), and model fit indices commonly used in confirmatory factor analysis (Clark & Bowles, 2018; Finch, 2020b). Sufficient evidence has not accumulated as yet to justify use of these methods. In contrast, there are several methods of determining the number of factors to retain that are *not* supported and should be ignored. These include the so-called "Eigenvalue 1" rule as well as the chi-square test with ML extraction (Flora, 2018; Hayashi et al., 2007; Russell, 2002; Velicer et al., 2000).

Summary of Empirical Guidelines

Velicer et al. (2000) recommended that a combination of PA and MAP should be employed, with scree reserved as a potentially useful adjunct. Other methodologists have also recommended PA and/or MAP (Bandalos, 2018; DeVellis, 2017; Fabrigar & Wegener, 2012; Fabrigar et al., 1999; Ford et al., 1986; Hair et al., 2019; Hayton et al., 2004; Hoelzle & Meyer, 2013; Howard, 2016; Hoyle & Duvall, 2004; Kanyongo, 2005; Lawrence & Hancock, 1999; Osborne, 2014; Sakaluk & Short, 2017), noting that PA tends to slightly overextract whereas MAP tends to slightly underextract.

Empirical Criteria with Stata

Stata can provide all three methods recommended by Velicer et al. (2000). However, Stata structures its EFA routines into two sequential components. First, EFA or PCA must be conducted to extract the factors or components. Second, postestimation methods such as scree, PA, and MAP are implemented. In this case, we are not interested in the EFA or PCA results, only the postestimation results. Consequently, the `quietly pca vocab1-designs2, components(8)` command will be used to suppress display of the extraction results.

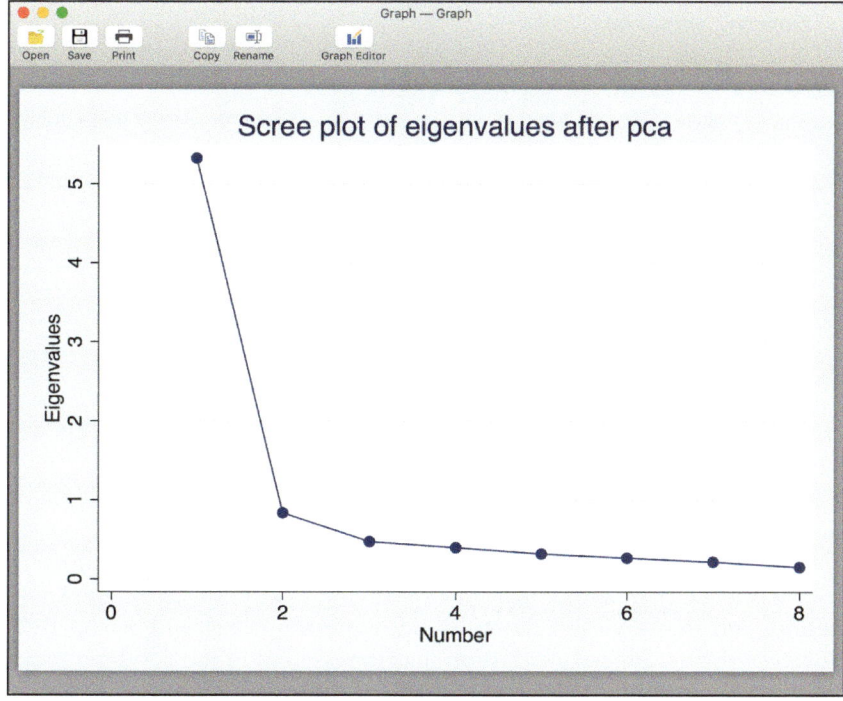

Figure 12.2 Scree plot of iq data in Stata graph window

Scree Plot

A scree plot can be accessed through the ***Statistics > Multivariate analysis > Factor and principal components analysis > Postestimation > Scree plot of eigenvalues*** menu options. Alternatively, the screeplot command will produce the same result (Figure 12.2). The scree plot for the iq data appears to favor the extraction of two factors.

Parallel Analysis

A user-contributed ado-file called fapara offers a postestimation PA option. It can be downloaded via the ssc install fapara command. The fapara, pca reps(500) postestimation command produces the results displayed in Figure 12.3. In this case, the first real eigenvalue (5.33) was larger than the first random eigenvalue (1.36) but the subsequent random eigenvalues were larger than the real eigenvalues, indicating that only one factor might be sufficient.

PA can also be conducted with online calculators available at www.statstodo .com/ParallelAnalysis_Exp.php and https://analytics.gonzaga.edu/parallelengine.

```
. fapara, pca reps(500)

PA -- Parallel Analysis for Principal Components -- N = 152
PA Eigenvalues Averaged Over 500 Replications
            PCA          PA          Dif
    1.    5.328035  >  1.35784    3.970195
    2.    .8370123  <  1.218015   -.3810031
    3.    .4743501     1.117071   -.6427205
    4.     .398908     1.028622   -.6297141
    5.    .3203761     .9491716   -.6287955
    6.    .2683994     .8650925   -.5966932
    7.    .2193804     .7803984    -.561018
    8.    .1535386     .6837893   -.5302507
```

Figure 12.3 Parallel analysis results from Stata fapara ado-file

Alternatively, a standalone program called *Monte Carlo PCA for Parallel Analysis* can be downloaded from edpsychassociates.com/Watkins3.html.

Parallel analysis can also be conducted with the eigenvalues extracted from a reduced correlation matrix by specifying a common factor extraction method and some researchers prefer this approach (Crawford et al., 2010). However, the unreduced correlation matrix was used in the development of parallel analysis (Horn, 1965) and in much of the simulation research (Velicer et al., 2000; Zwick & Velicer, 1986) and has been found more accurate than results from the reduced matrix (Auerswald & Moshagen, 2019; Garrido et al., 2013; Lim & Jahng, 2019). Additionally, the unreduced matrix was the foundation for development of both MAP and scree so it seems reasonable to "use it to determine the spread of variance across the factors and as the basis for deciding on the number of factors to be extracted for the next stage" (Child, 2006, p.153).

Minimum Average Partial

A user-contributed ado-file called minap offers a postestimation MAP option. It can be downloaded via the `ssc install minap` command and implemented by the `minap vocab1-designs2` postestimation command to produce the results displayed in Figure 12.4. The MAP values sequentially drop from .0697 for one component to .0626 for two components and then increase to .0949 for three components. The lowest MAP value identifies the number of factors to retain. In this case, MAP reached a minimum at two components.

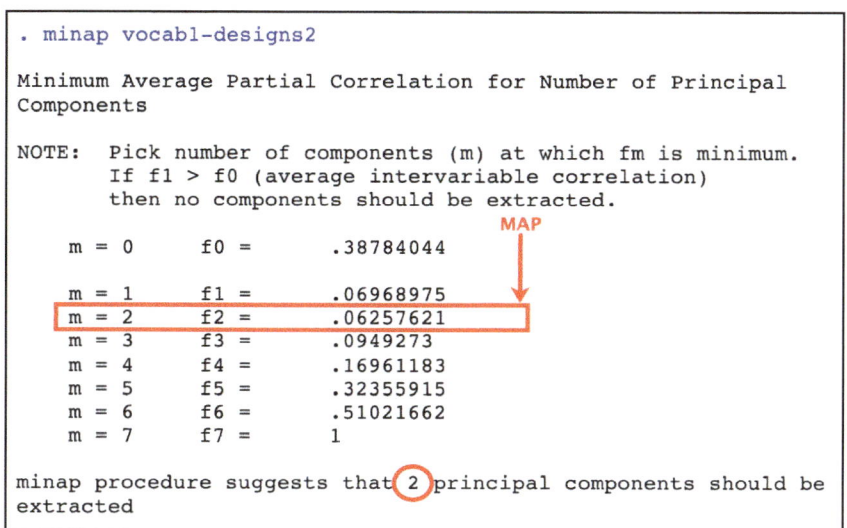

Figure 12.4 Minimum average partial results from Stata minap ado-file

Theoretical Knowledge

Although empirical guidelines are useful, there is no guarantee that they are correct (Bandalos, 2018; Cattell, 1966; Fabrigar & Wegener, 2012; Gorsuch, 1983; Pett et al., 2003; Preacher et al., 2013; Widaman, 2012). The accuracy of empirical guidelines is more likely to be compromised when factors are highly correlated, factor loadings are low, the number of factors is large, and the sample size is small (Lim & Jahg, 2019). Given this fundamental uncertainty, methodologists have recommended that multiple criteria, including relevant theory and previous research, be used to determine the number of factors to retain (Bandalos, 2018; Bandalos & Boehm-Kaufman, 2009; Bandalos & Finney, 2019; Brown, 2015; Fabrigar et al., 1999; Finch, 2013, 2020a, 2020b; Flora, 2018; Hoelzle & Meyer, 2013; McCoach et al., 2013; Norman & Streiner, 2014; Nunnally & Bernstein, 1994; Osborne, 2014; Pituch & Stevens, 2016; Preacher & MacCallum, 2003; Preacher et al., 2013; Reio & Shuck, 2015; Velicer & Fava, 1998; Widaman, 2012). There is no prior research with these eight measured variables, but research with similar verbal and nonverbal measures have suggested two factors and development of these tests was guided by that theoretical expectation.

Model Selection

Given that there is no infallible method to identify the "true" number of factors, Cattell (1978) said that, "taking out the right number of factors does not mean in most cases a number correct in some absolute sense but in the sense of not missing any factor of more than trivial size" (p. 61). This suggests a strategy of "selecting from among a set of competing theoretical explanations the model that best balances the desirable characteristics of parsimony and fit to observed data" (Preacher et al., 2013, p. 29). Each candidate model contains a different number of factors and is judged on its interpretability and conceptual sense in this model selection process (Bandalos, 2018; Carroll, 1993; Cudeck, 2000; Fabrigar & Wegener, 2012; Fabrigar et al., 1999; Finch, 2013; Flora, 2018; Ford et al., 1986; Gorsuch, 1983, 1988, 1997; Hair et al., 2019; Hoelzle & Meyer, 2013; Kahn, 2006; McCoach et al., 2013; Nunnally & Bernstein, 1994; Osborne, 2014; Osborne & Banjanovic, 2016; Pituch & Stevens, 2016; Preacher & MacCallum, 2003; Preacher et al., 2013; Schmitt et al., 2018; Tabachnick & Fidell, 2019; Velicer et al., 2000; Widaman, 2012). Of course, a model that is generalizable to other samples is scientifically desirable but multiple samples and multiple EFAs may be required to achieve that goal (Preacher et al., 2013).

Report

Velicer et al. (2000) recommended that a combination of parallel analysis (PA; Horn, 1965) and minimum average partial (MAP; Velicer, 1976) methods should be employed for determining the number of factors to retain for rotation, with scree (Cattell, 1966) as a potentially useful adjunct. Using these three criteria, it appeared that one or two factors would be sufficient for an optimal balance between comprehensiveness and parsimony. Two factors were also signaled as sufficient by prior research and theory. To ensure that underextraction did not occur, a model with three factors might also be considered (Cattell, 1978). Therefore, models with three, two, and one factor(s) will be sequentially evaluated for their interpretability and theoretical meaningfulness.

13 Step 8

Rotate Factors

Exploratory factor analysis (EFA) extraction methods have been mathematically optimized to account for the covariance among measured variables but they do not take interpretability into account (Bandalos, 2018; Comrey & Lee, 1992; DeVellis, 2017; Nunnally & Bernstein, 1994; Pituch & Stevens, 2016). As previously described, the orientation of the factor axes is fixed in factor space during factor extraction. Those axes can be rotated about their origin to make the loadings more interpretable without changing the underlying structure of the data (Tabachnick & Fidell, 2019). The rotated structure accounts for the same proportion of variance as the unrotated structure but it distributes that variance across factors differently (Sheskin, 2011). Typically, variance from the first factor is redistributed to the following factors.

For example, extraction of only two factors from the iq data found that the first and second factors accounted for 62.3% and 7.3% of the total variance, respectively, before rotation. However, 38.8% of that variance was allocated to the first factor and 30.8% to the second factor following an orthogonal rotation. Thus, the total variance of 69.6% was distributed differently between unrotated and rotated factors. Rotation is akin to taking a photograph of a person from multiple angles. The person remains the same but some of the photographs better represent the person than others. DeVellis (2017, pp. 171–176) described several other analogies to help understand the concept of rotation.

As illustrated with the iq data, the first unrotated factor accounts for the largest amount of explained variance among the measured variables. However, the first rotated factor will not necessarily account for the maximum variance following rotation because all rotation methods redistribute the variance. "Indeed, an extremely common error in the published literature involves researchers erroneously interpreting the eigenvalues associated with the unrotated factors as reflecting the variance accounted for by the factors after rotation" (Hetzel, 1996, p. 192). A recent example of such an erroneous interpretation was described by Dombrowski et al. (2021).

For a simple example of rotation, the two factor initial (unrotated) solution for the iq data was obtained and conceptually plotted on a two-dimensional graph in factor space in Figure 13.1. It is apparent that the variables tend to

DOI: 10.4324/9781003149286-13

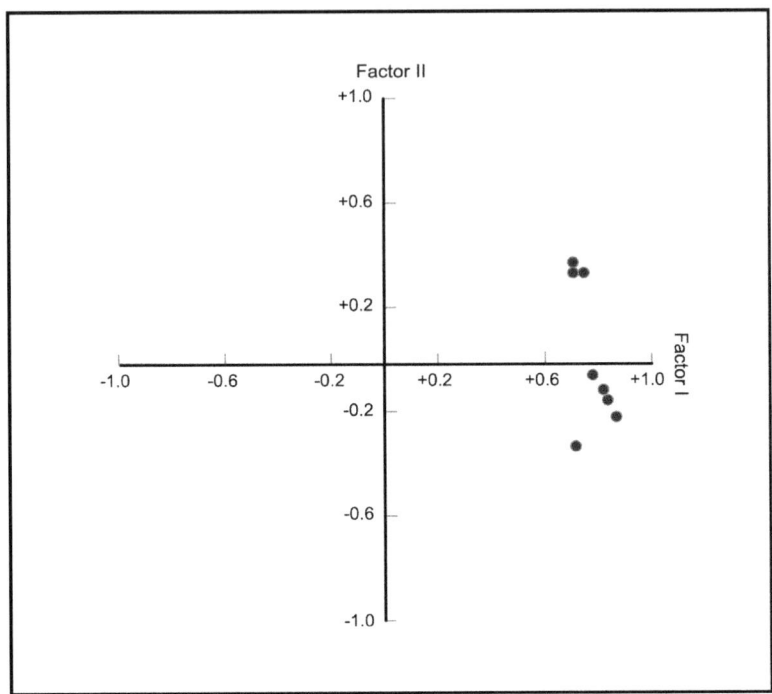

Figure 13.1 Unrotated factor plot

form two clusters in multidimensional space but they are not particularly well aligned with the factor axes.

Perhaps the factor axes could be rotated about their origin to bring them closer to the center of the variable clusters (Osborne & Banjanovic, 2016)? As displayed in gray in Figure 13.2, the rotated Factor II axis is now closer to the cluster of three variables in factor space but that has moved the axis of Factor I away from the second variable cluster. This is called an orthogonal rotation because the factor axes have been restrained to right angles. Due to this restraint, the two factors also remain uncorrelated because the cosine of 90° is zero.

Alternatively, the factor axes could be rotated independent of each other to bring each of them closer to the center of the variable clusters. That is, rotate the axes without constraining them to orthogonality. Independent rotation of the axes is called oblique rotation. This allows the factors to become correlated if that results in improved simplicity of loadings. As displayed in gray in Figure 13.3, that has brought both axes closer to the variable clusters. The angle between the factor axes is approximately 45° and the correlation between the two factors is estimated by the cosine of that angle (i.e., ~.71).

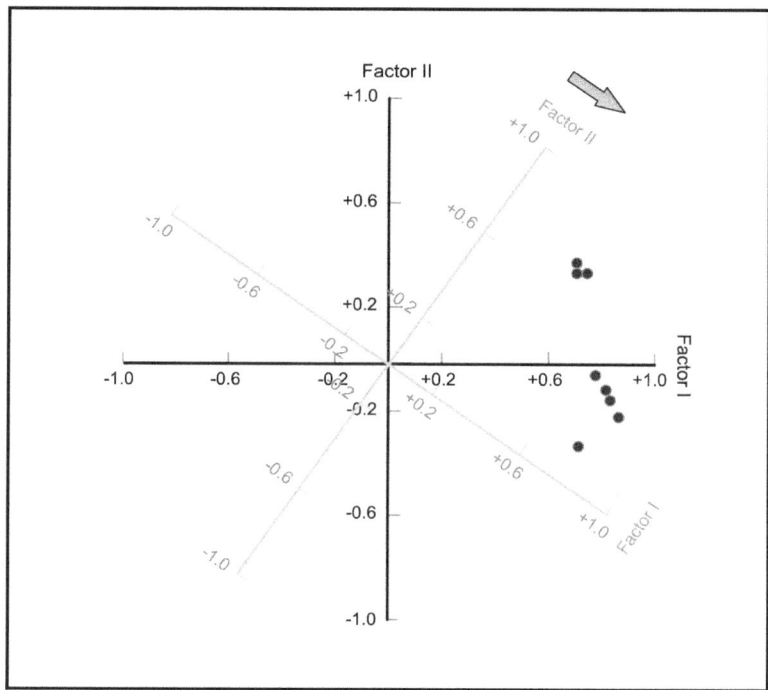

Figure 13.2 Orthogonally rotated factor plot

These geometric representations of factor rotation have been only approximate, sufficient for illustration of the concept of rotation but not precise enough for use in practice. They also intimate that "there is an infinite number of equally fitting ways those factors may be oriented in multidimensional space" (Sakaluk & Short, 2017, p. 3), which is called rotational indeterminacy (Fabrigar & Wegener, 2012). Rotational indeterminacy is an additional reason that methodologists stress that factor models must be judged on their interpretability and theoretical sense (Bandalos, 2018; Cudeck, 2000; Fabrigar & Wegener, 2012; Fabrigar et al., 1999; Ford et al., 1986; Gorsuch, 1983, 1988; Hair et al., 2019; Kahn, 2006; Nunnally & Bernstein, 1994; Osborne, 2014; Preacher & MacCallum, 2003; Preacher et al., 2013; Tabachnick & Fidell, 2019; Velicer et al., 2000; Widaman, 2012).

Orthogonal Versus Oblique Rotation

Researchers differ on their preference for orthogonal or oblique rotations. Those who prefer orthogonal rotations cite simplicity and ease of interpretation (Mertler & Vannatta, 2001). However, orthogonal rotations are

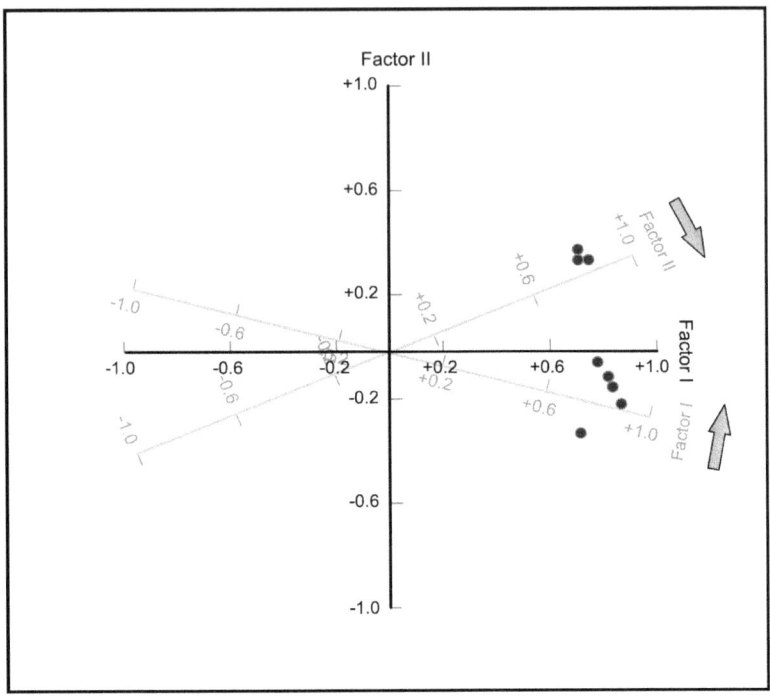

Figure 13.3 Obliquely rotated factor plot

inappropriate when there is a higher-order factor (see the chapter on higher-order and bifactor models) because they disperse the general factor variance across the first-order factors thereby artificially obscuring the higher-order factor (Gorsuch, 1983). Other researchers prefer oblique rotations due to their accuracy and to honor the reality that most variables are correlated to some extent (Bandalos & Boehm-Kaufman, 2009; Bandalos & Finney, 2019; Brown, 2013, 2015; Costello & Osborne, 2005; Cudeck, 2000; Fabrigar & Wegener, 2012; Fabrigar et al., 1999; Finch, 2013; Flora, 2018; Flora et al., 2012; Ford et al., 1986; Gorsuch, 1983; McCoach et al., 2013; Meehl, 1990; Mulaik, 2010, 2018; Osborne, 2014; Pett et al., 2003; Pituch & Stevens, 2016; Preacher & MacCallum, 2003; Reio & Shuck, 2015; Reise et al., 2000; Rummel, 1967; Russell, 2002; Sakaluk & Short, 2017; Sass, 2010; Widaman, 2012; Worthington & Whittaker, 2006; Zhang & Preacher, 2015).

Some researchers suggest using oblique rotation only if the correlation between factors exceeds .20 (Finch, 2020a) or .32 (Roberson et al., 2014; Tabachnick & Fidell, 2019), whereas others believe there is no reason to select one type of rotation over another so both should be applied (Child, 2006; Hair et al., 2019). Nonetheless, the arguments in favor of oblique rotations are compelling. As articulated by Schmitt (2011), "because oblique rotation methods generally produce accurate and comparable factor structures to orthogonal methods even when interfactor correlations are negligible, it is strongly recommend that researchers only use oblique rotation methods because they generally result in more realistic and more statistically sound factor structures" (p. 312).

In practice, rotation is accomplished with algebraic algorithms (analytic rotations). The most prominent orthogonal rotation is varimax (Kaiser, 1958). Others include quartimax, equamax, and parsimax. Stata includes varimax, quartimax, equamax, parsimax, and entropy orthogonal rotation options. The most popular oblique rotations are promax (Hendrickson & White, 1964) and oblimin (Jennrich & Sampson, 1966). Other oblique rotations include geomin, maxplane, orthoblique, direct quartimin, bigquartmin, promin, and covarimin. Stata offers promax, oblimin, oblimax, quartimin, and target oblique rotation options. Technical reviews of analytic rotations are available if further details are desired (Browne, 2001; Sass & Schmitt, 2010; Schmitt & Sass, 2011). In many cases, different rotations within the orthogonal and oblique families are likely to produce similar results (Bandalos & Finney, 2019; Nunnally & Bernstein, 1994; Sass & Schmitt, 2010).

Many methodologists have recommended that varimax be selected if an orthogonal rotation is employed (Child, 2006; Gorsuch, 1983, 2003; Kline, 1994; Nunnally & Bernstein, 1994). It is not clear whether there is a superior oblique method (Schmitt & Sass, 2011). Methodologists have variously recommended oblimin (Child, 2006; Flora, 2018; McCoach et al., 2013; Mulaik, 2018), geomin (Hattori et al., 2017), and promax (Finch, 2006; Goldberg & Velicer, 2006; Gorsuch, 1983, 1988, 2003; Matsunaga, 2010; Morrison, 2009; Russell, 2002; Sass, 2010; Thompson, 2004) and those methods appear to be most common in practice (Goretzko et al., 2019). Given this choice among oblique rotations, many methodologists recommend that results from another rotation method be compared to the initial choice to ensure that results are robust to rotation method (Finch, 2020a; Hattori et al., 2017). Promax and oblimin would be the obvious choice for this comparison.

Oblique Parameters

Oblique rotations require that a parameter be set that specifies the extent to which the factors are allowed to be correlated. That parameter is referred to as delta in oblimin, k or kappa in promax, and ε in geomin. Oblimin and geomin solutions seem to be very sensitive to the parameter setting whereas promax is relatively insensitive to the k setting (Gorsuch, 2003; Tataryn et al., 1999). For oblimin rotations, Howard (2016) noted that delta values of zero were preferred, which is the default value in Stata. Users can apply an alternative delta value by including it in the rotation command: `rotate, oblimin(-1)`. For geomin rotations, ε values of .01 and .05 were recommended by Hattori et al. (2017) and Morin et al. (2020), respectively. Unfortunately, geomin may converge on solutions that are not optimal

(Hattori et al., 2017). For promax rotations, kappa values of two through four will probably be adequate, although Stata's default value of three is probably best for the version of promax it applies (Gorsuch, 2003; Tataryn et al., 1999). Users can apply an alternative kappa value by including it in the rotation command: `rotate, promax(4)`.

These parameters can be used to fine-tune an oblique solution through the use of a hyperplane count method (Gorsuch, 1983). That is, multiple rotation solutions can be performed, each with a different parameter value, and the parameter value that produces the solution with the largest number of essentially zero loadings can be selected. Although the exact definition of "essentially zero" is subjective, .05 or .10 have generally been used (Gorsuch, 1983). Cattell (1952, 1978) often used the percentage of the elements in the factor loading matrix that are in the hyperplane (i.e., −.10 to +.10) as his analytic criterion, but use of the hyperplane count has received little attention in recent years.

Factor Loadings

To this point, factor loadings have been specified as quantifications of the relationship between measured variables and factors. By allowing factors to correlate, oblique rotations produce two types of factor loadings: pattern coefficients and structure coefficients. Pattern coefficients quantify the relationship between a measured variable and its underlying factor, after the effects of the other variables have been taken into account or "partialled out". Pattern coefficients are regression-like weights and may occasionally be greater than ± 1.00. In contrast, structure coefficients reflect the simple correlation of a measured variable with its underlying factor and ignore the effects of other factors and must, therefore, range from −1.00 to +1.00. Consequently, it is no longer sufficient to refer to factor loadings after an oblique rotation has been applied. Rather, the appropriate coefficient *must* be identified (pattern or structure).

Pattern and structure coefficients will be identical if the factors are perfectly uncorrelated with each other and will increasingly deviate from each other as the factor intercorrelations increase. Methodologists typically recommend that both pattern and structure coefficients should be interpreted (Gorsuch, 1983; Matsunaga, 2010; McClain, 1996; McCoach et al., 2013; Nunnally & Bernstein, 1994; Pett et al., 2003; Reio & Shuck, 2015; Thompson, 2004) because that "increases the interpretative insights gained from a factor analysis" (Hetzel, 1996, p. 183). However, pattern coefficients and factor intercorrelations should receive primary attention during the model evaluation process (Bandalos, 2018; Brown, 2015; Cattell, 1978; Fabrigar & Wegener, 2012; Flora, 2018; Gibson et al., 2020; Hair et al., 2019; McCoach et al., 2013; Mulaik, 2010; Pituch & Stevens, 2016; Rencher & Christensen, 2012; Roberson et al., 2014; Sakaluk & Short, 2017). Comrey and Lee (1992) suggested that pattern coefficients of .70 are excellent, .63 are very good, .55 are good, .45 are fair, and .32 are poor. Other methodologists have suggested that loadings ≥ .50 are satisfactory (Goldberg & Velicer, 2006; Morin et al., 2020). Structure coefficients can be consulted after the final model has been tentatively accepted to ensure that the factors have been named appropriately and that anomalous results have not been inadvertently accepted (Kahn, 2006).

Eigenvalues and Variance Extracted

The unrotated and rotated factor solutions will explain the same amount of total variance (computed with eigenvalues) but rotation spreads that variance across the factors to improve interpretation and parsimony. Given this redistribution of variance across factors, it is necessary to identify whether the unrotated or rotated solution is referenced when considering the proportion of total variance that was attributed to each factor. Given that rotation apportions variance away from the first factor to later factors, it would be inappropriate to interpret the proportion of variance following rotation as an indicator of factor importance (Hetzel, 1996).

Report

An oblique rotation was selected because it honors the ubiquity of intercorrelations among social science variables (Meehl, 1990). Among the potential oblique analytic rotations, promax was chosen because it is an oblique modification of the widely accepted varimax procedure (Gorsuch, 1983; Thompson, 2004). To ensure stability across extraction methods, oblimin extraction was also employed (Finch, 2020a).

14 Step 9

Interpret Exploratory Factor Analysis Results

Model Selection Guidelines

Exploratory factor analysis (EFA) models with different numbers of factors should be sequentially evaluated for their interpretability and theoretical meaningfulness (Fabrigar & Wegener, 2012; Finch, 2020a; Flora, 2018). There are a variety of guidelines that can be used to judge models. It is crucial that the researcher explicitly detail the judgment guidelines that will be applied *prior* to implementation. This a priori explanation will reduce the possibility of self-serving judgments (Rubin, 2017; Simmons et al., 2011). Following is an enumerated list of guidelines that are most likely to be helpful.

1. Establish a threshold at which factor loadings (pattern coefficients for oblique rotations) will be considered meaningful (Worthington & Whittaker, 2006). Conventionally, loadings that meet this threshold are characterized as *salient*. It is common to arbitrarily consider factor loadings of .30, .32, or .40 as salient (Child, 2006; Comrey & Lee, 1992; Hair et al., 2019; Pituch & Stevens, 2016). That is, variables with around 9%, 10%, or 16% (loading squared) of their variance explained by the factor. Some researchers consider .30 or .32 salient for EFA and .40 salient for PCA. These thresholds honor practical significance but ignore statistical significance. That is, a loading of .32 might account for 10% of a variable's variance but it might not be statistically significantly different from zero, thereby calling into question its stability (Schmitt & Sass, 2011; Zhang & Preacher, 2015). Norman and Streiner (2014) suggested an approximation based on Pearson correlation coefficients to compute the statistical significance ($p = .01$) of factor loadings: $\dfrac{5.152}{\sqrt{N-2}}$. For the iq data, statistical significance ($p = .01$) would equate to $5.152 \div 12.25 = .42$. A more relaxed $p = .05$ standard would modify the numerator: $\dfrac{3.92}{\sqrt{N-2}}$ or $3.92 \div 12.25 = .32$. Thus, establish a threshold for salience that is both practically and statistically significant *before* conducting EFA.
2. Respect the concept of simple structure (Thurstone, 1947). Variables with salient loadings on more than one factor are said to cross-load

DOI: 10.4324/9781003149286-14

and are called factorially complex variables (Sheskin, 2011). Complex loadings might be appropriate for some structures, but will complicate interpretation. Simple structure solutions will probably be more interpretable and more likely to replicate. Conceptually, simple structure implies that each variable will exhibit salient loadings on a few factors (the fewer the better) and weak loadings on all other factors (Pituch & Stevens, 2016). At its simplest, several variables will saliently load onto each factor and each variable will saliently load onto only one factor. In practice, the goal is a reasonable approximation of simple structure (Morin et al., 2020). Simple structure recognizes "the purpose of science [which] is to uncover the relatively simple deep structure principles or causes that underlie the apparent complexity observed at the surface structure level" (Le et al., 2010, p. 112) and embodies the scientific principle of parsimony (Harman, 1976).

3. If the measured variables are items or are otherwise meant to be combined into a scale, the alpha reliability (Cronbach, 1951) of each factor should exceed the threshold established for the use of such scales (Pett et al., 2003). For example, reliability coefficients in the .90s are likely sufficient for clinical decisions (DeVellis, 2017), coefficients in the .80s are satisfactory for non-critical decisions, coefficients in the .70s are adequate for group experimental research, and coefficients less than .70 are inadequate for most applications (Hunsley & Mash, 2007; Kline, 2013).

4. Measures of model fit, including:

   a. Residuals. The difference between the actual correlation matrix and a correlation matrix reproduced by the model. The average overall residual misfit is quantified by the root mean squared residual (RMSR). The smaller the RMSR value the better, with values ≤ .08 preferred (Brown, 2015). RMSR values will continue to decrease as more factors are extracted so several models may exhibit RMSR values ≤ .08. The goal is to select the model where RMSR is substantially smaller than a model with one more factor but does not appreciably decrease when another factor is removed.

   Individual residual correlations should also be considered. Ideally, the proportion of non-redundant residual correlations greater than the absolute value of .05 should be small (Finch, 2020a; Garson, 2013; Johnson & Morgan, 2016; Maydeu-Olivares, 2017), with absolute residuals > .10 more strongly indicating the presence of another factor (Cudeck, 2000; Flora, 2018; Kline, 2013; McDonald, 2010). If the residuals "are not both small and without apparent pattern, additional factors may be present in the data" (Nunnally & Bernstein, 1994, p. 471). As summarized by Flora (2018), "although there is no concrete guideline or cut-off for how small residual correlations should be, I suggest that any residual correlation > .10 is worthy of further consideration with respect to potential model misfit" (p. 255). Maydeu-Olivares (2017) found that RMSRs from samples tended to be higher than RMSRs from populations and suggested that the standard of close fit should be an RMSR value of .05 with no individual residual larger than .10.

b. Bayesian information criterion (BIC; Schwarz, 1978). An index that balances model simplicity versus goodness of fit. There is no absolute good or bad BIC value: the model with the lowest BIC value is preferred. The BIC was designed to detect the "true" model if it is among the set of candidate models and emphasizes parsimony by including a penalty for model complexity (Burnham & Anderson, 2004). Unfortunately, the BIC tends to overestimate the number of factors as sample size increases (Schmitt et al., 2018), has received little research attention for use with EFA models, and is only computed by Stata with ML estimation.

c. Indices of model fit used in CFA, including the comparative fit index (CFI), Tucker-Lewis index (TLI), root mean square error of approximation (RMSEA), etc. Similar to the BIC, these indices have received little research attention in the EFA context (Montoya & Edwards, 2021). However, one study found that they were of "questionable utility" for determining the number of factors in EFA (Clark & Bowles, 2018, p. 544) although CFI/TLI values ≥ .95 might protect against underfactoring. Another study found that RMSEA difference values ≥ .015 might be helpful in determining the number of factors to retain (Finch, 2020b). Other studies found that all fit indices were influenced by properties of the data and model, making them less accurate than parallel analysis (Garrido et al., 2016; Montoya & Edwards, 2021). No indices of model fit are provided for EFA by Stata.

5. Symptoms of model misfit due to overfactoring:

   a. Factors with only one (singlet) or two (doublet) salient loadings. Such factors are relatively weak and unlikely to replicate (Bandalos, 2018; Bandalos & Finney, 2019; Benson & Nasser, 1998; Brown, 2015; Fabrigar & Wegener, 2012; Nunnally & Bernstein, 1994; Preacher & MacCallum, 2003; Velicer & Fava, 1998). Factors with at least three salient loadings are preferred (Comrey & Lee, 1992; Garson, 2013; Johnson & Morgan, 2016; Mulaik, 2010, 2018; Reio & Shuck, 2015; Velicer & Fava, 1998) because "no meaningful component can be identified unless each factor is overdetermined with three or four or more tests" (Thurstone, 1937, p. 75). Singlet and doublet variables will likely exhibit low communality (Fabrigar & Wegener, 2012) and therefore little explanatory power.

   b. Factors that are very highly correlated. Interfactor correlations that exceed .80 or .85 might be a sign of overfactoring and pose a threat to discriminant validity (Brown, 2015; Finch, 2020a; McCoach et al., 2013; Schmitt et al., 2018), whereas interfactor correlations > .90 probably mean that "the two factors are clearly not distinct" (Kline, 2013, p. 185).

    c.   Factors based on similarities in variable distributions rather than similarity of variable content (Bandalos, 2018).

    d.   Unreasonable parameter estimates called Heywood cases (e.g., communalities > 1.00), nonpositive definite matrices, or a failure of iterations to converge on a factor extraction solution may indicate a misspecified factor model (Flora, 2018; Lorenzo-Seva & Ferrando, 2021; Pituch & Stevens, 2016; Wothke, 1993).

    e.   Factor splitting. Measured variables that are known to load on a single factor in the population are split onto multiple factors after rotation (Wood et al., 1996).

6.   Symptoms of model misfit due to underfactoring:

    a.   Measured variables that saliently load on a factor do not seem to reflect a common unifying theme. This may indicate that two or more factors have collapsed onto the same factor (Fabrigar & Wegener, 2012).

    b.   Poor model fit indices and modest loadings of measured variables on all factors or many complex loadings (Benson & Nasser, 1998; Zhang, 2014).

7.   Robustness of results across extraction and rotation methods. As suggested by Gorsuch (1983), "factor the data by several different analytic procedures and hold sacred only those factors that appear across all the procedures used" (p. 330).

Report

Models with three, two, and one factor(s) will be sequentially evaluated for their interpretability and theoretical meaningfulness using several guidelines. Given oblique rotation, pattern coefficients and factor intercorrelations will receive primary attention during the model evaluation process (Bandalos, 2018; Hair et al., 2019). To ensure both practical (10% variance explained) and statistical ($p < .05$) significance of pattern loadings, the threshold for salience will be set at .32 (Norman & Streiner, 2014) with a goal of approximate simple structure (Morin et al., 2020; Thurstone, 1947). The models will be compared on average model misfit (RMSR) and the proportion of residual coefficients that exceed absolute values of .05 and .10 (Flora, 2018; Nunnally & Bernstein, 1994). Close model fit will be indicated by an RMSR value of .05 or smaller and no individual residual coefficient greater than .10 (Maydeu-Olivares, 2017). Additionally, the alpha reliability of scales created from the salient variables of each factor should approach .90, given the intended clinical use of these scales (DeVellis, 2017). Finally, each model will be examined for symptoms of overextraction such as statistically improper solutions, fewer than three salient loadings, technical factors, etc. and symptoms of underextraction such as no common unifying theme, many complex loadings, etc. (Bandalos, 2018; Fabrigar & Wegener, 2012).

Model Evaluation

The first model to be evaluated will be illustrated with both menu and command code for users who prefer either approach. Thereafter, only commands will be demonstrated. Given the many objects created by Stata during the EFA procedure, objects will be presented individually and sequentially rather than in total.

Model 3

Access to EFA via the Stata menu system begins with ***Statistics > Multivariate analysis > Factor and principal component analysis > Factor analysis*** options as illustrated in Figure 14.1.

After selecting the Factor analysis option, the **factor – Factor Analysis** window opens where the variables to be used in the factor analysis model must be specified as in Figure 14.2.

After selecting the eight measured variables (vocab1–designs2), the extraction method and stopping rule for extractions must be specified by selecting the ***Model 2*** tab. In this case, ML estimation and three factors are stipulated as in Figure 14.3.

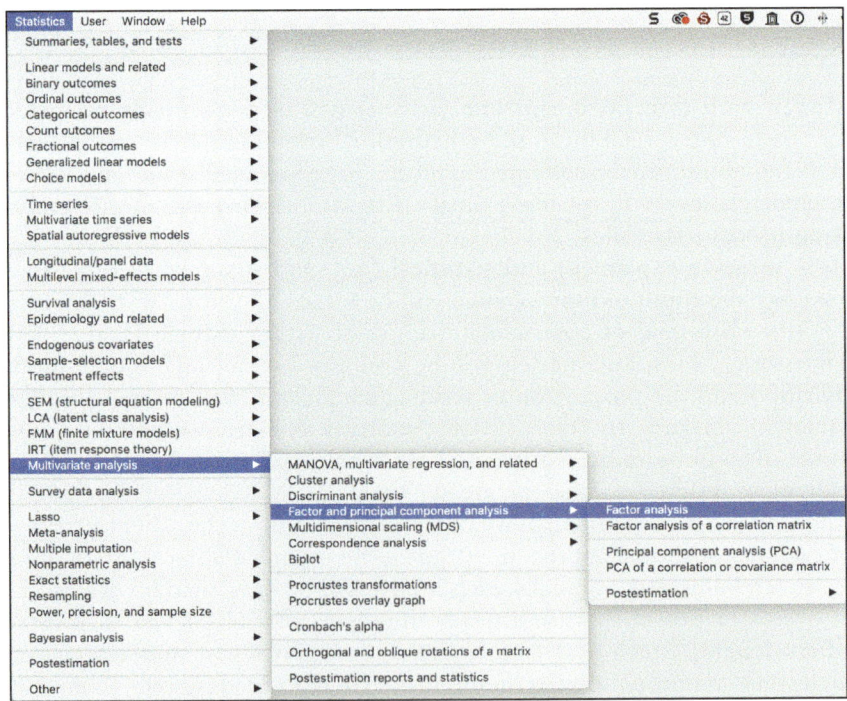

Figure 14.1 Stata multivariate statistics menu system

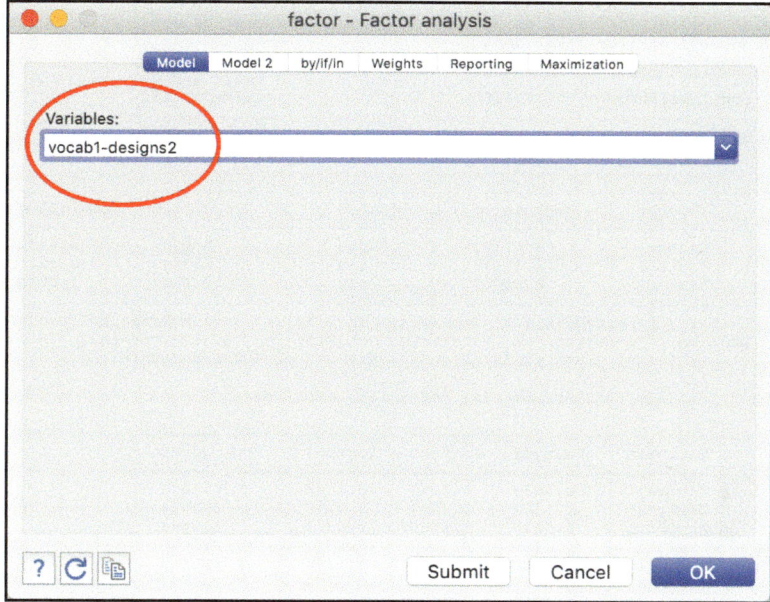

Figure 14.2 Stata factor analysis model window

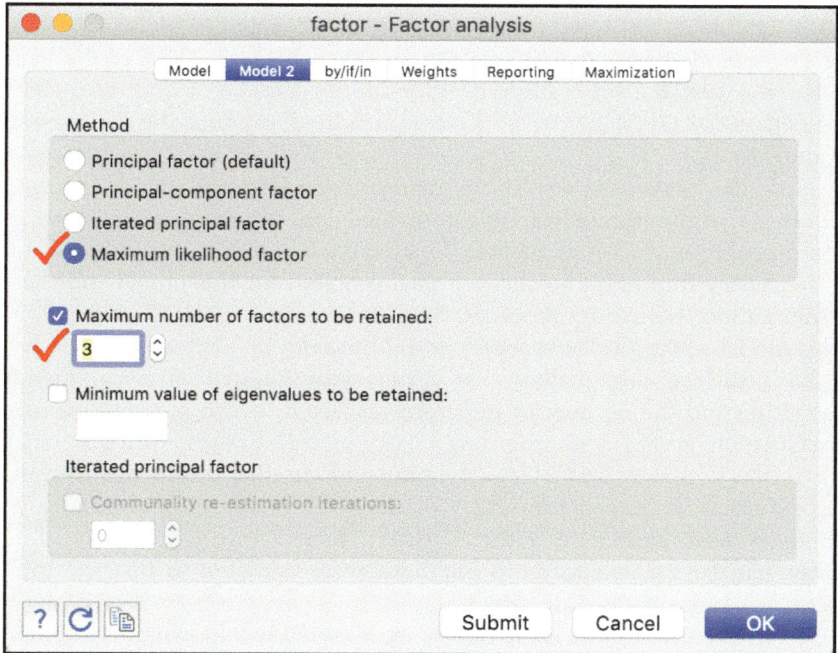

Figure 14.3 Stata factor analysis model2 window

```
Factor analysis/correlation                  Number of obs    =        152
   Method: maximum likelihood                Retained factors =          3
   Rotation: (unrotated)                     Number of params =         21
                                             Schwarz's BIC    =    116.881
   Log likelihood = -5.689615                (Akaike's) AIC   =    53.3792

   Beware: solution is a Heywood case
           (i.e., invalid or boundary values of uniqueness)

   ----------------------------------------------------------------------
       Factor  |  Eigenvalue   Difference       Proportion   Cumulative
   ----------+-----------------------------------------------------------
      Factor1  |    3.65377     1.74680            0.6072       0.6072
      Factor2  |    1.90698     1.45040            0.3169       0.9241
      Factor3  |    0.45657        .               0.0759       1.0000
   ----------------------------------------------------------------------
   LR test: independent vs. saturated:   chi2(28) =  893.09 Prob>chi2 = 0.0000
   LR test:   3 factors vs. saturated:   chi2(7)  =   10.97 Prob>chi2 = 0.1400
   (tests formally not valid because a Heywood case was encountered)

Factor loadings (pattern matrix) and unique variances

   ------------------------------------------------------------
      Variable |  Factor1    Factor2    Factor3  | Uniqueness
   ----------+---------------------------------+----------------
        vocab1 |   0.5637     0.7253    -0.1044  |   0.1452
      designs1 |   0.5862     0.4027     0.4719  |   0.2715
      similar1 |   0.5818     0.6260    -0.0362  |   0.2684
       matrix1 |   0.7146     0.3031     0.2371  |   0.3413
      veranal2 |   0.6480     0.4571    -0.1268  |   0.3550
        vocab2 |   0.5773     0.6605    -0.1350  |   0.2122
       matrix2 |   1.0000    -0.0000    -0.0000  |   0.0000
      designs2 |   0.6244     0.2996     0.3622  |   0.3891
```

Figure 14.4 Unrotated solution for three factors

The specified EFA will be computed and reported in the **Results** pane when the **OK** button is clicked. Alternatively, the `factor vocab1-designs2, ml factors(3)` command will produce the same results (Figure 14.4).

Note that Stata gave a warning message regarding a Heywood case and reported that the matrix2 variable contained zero unique variance. Thus, its communality was 1.00 or greater. Technically, a communality value > 1.0 is statistically impossible so this is a fatal problem for this model and these results cannot be trusted (Lorenzo-Seva & Ferrando, 2021; Wothke, 1993). This model is statistically improper and should not be interpreted.

Nevertheless, this analysis was rerun with iterated principal factors extraction and the number of iterations restricted to two as recommended by Gorsuch (1988) with the `factor vocab1-designs2, ipf factors(3) citerate(2)` command. In that alternative analysis, three factors accounted for 72.2% of the total variance, all communality estimates were legitimate, and the pattern matrix revealed that the third factor had two salient loadings but both were complex so that the third factor was not uniquely identified.

It was hypothesized that the three-factor model would exhibit symptoms of overextraction. The inadmissible results from a ML extraction and the failure to identify a third factor from a PF extraction were consistent with that hypothesis. Thus, there is no support for a model with three factors.

Model 2

As with the prior model, access to EFA options with the Stata menu system is via **Statistics > Multivariate analysis > Factor and principal component analysis > Factor analysis** options as illustrated in Figures 14.1 through 14.3 and retaining two factors. The same results can be obtained from the `vocab1-designs2, ml factors(2)` command.

Stata does not automatically display the communalities but they can be calculated by subtracting each uniqueness from one as shown in Figure 14.5. For this two-factor oblique solution communalities ranged from .598 to .855 with no inadmissible estimates. Communalities ≥ .60 are often considered to be high (Gibson et al., 2020). Alternatively, communalities can be computed with the `matrix M = J(1,colsof(e(Psi)),1) - e(Psi)` command and displayed with the `matrix list M` command.

The first factor accounted for 62.3% of the total variance (e.g., 4.984 ÷ 8.0) before rotation and the second factor added another 7.3% (e.g., 0.583 ÷ 8.0). Thus, two factors accounted for 70% of the total variance before rotation. This is a vivid demonstration of parsimony: two factors explained 70% of the variance of eight measured variables. The sum of the eigenvalues (4.984 + 0.583) is 5.567, which can be used to compute the relative contribution of each factor in terms of common variance. For example, 4.984 ÷ 5.567 = 89.53%. Stata reports these proportions (89.5% and 10.5%) for the first two factors before rotation.

```
Factor analysis/correlation                Number of obs    =       152
   Method: maximum likelihood              Retained factors =         2
   Rotation: (unrotated)                   Number of params =        15
                                           Schwarz's BIC    =    107.59
   Log likelihood = -16.11566              (Akaike's) AIC   =   62.2313

-----------------------------------------------------------------------
                                                   Common Variance
        Factor  |  Eigenvalue   Difference      Proportion   Cumulative
-----------------+-----------------------------------------------------
        Factor1  |    4.98449      4.40166          0.8953       0.8953
        Factor2  |    0.58283            .          0.1047       1.0000
-----------------------------------------------------------------------
   LR test: independent vs. saturated:  chi2(28) = 893.09 Prob>chi2 = 0.0000
   LR test:    2 factors vs. saturated: chi2(13) =  31.21 Prob>chi2 = 0.0031

Factor loadings (pattern matrix) and unique variances

        ---------------------------------------------------       Communality
          Variable |  Factor1   Factor2  |  Uniqueness |        (1 - uniqueness)
        -----------+--------------------+--------------+
             vocab1 |   0.8920   -0.2435 |     0.1451             .855
           designs1 |   0.7112    0.3036 |     0.4020             .598
           similar1 |   0.8472   -0.1140 |     0.2693             .731
            matrix1 |   0.7585    0.3361 |     0.3118             .688
           veranal2 |   0.7867   -0.0454 |     0.3791             .621
             vocab2 |   0.8613   -0.2125 |     0.2130             .787
            matrix2 |   0.7467    0.3450 |     0.3234             .677
           designs2 |   0.6868    0.3731 |     0.3892             .611
        ---------------------------------------------------
```

Figure 14.5 Unrotated solution for two factors

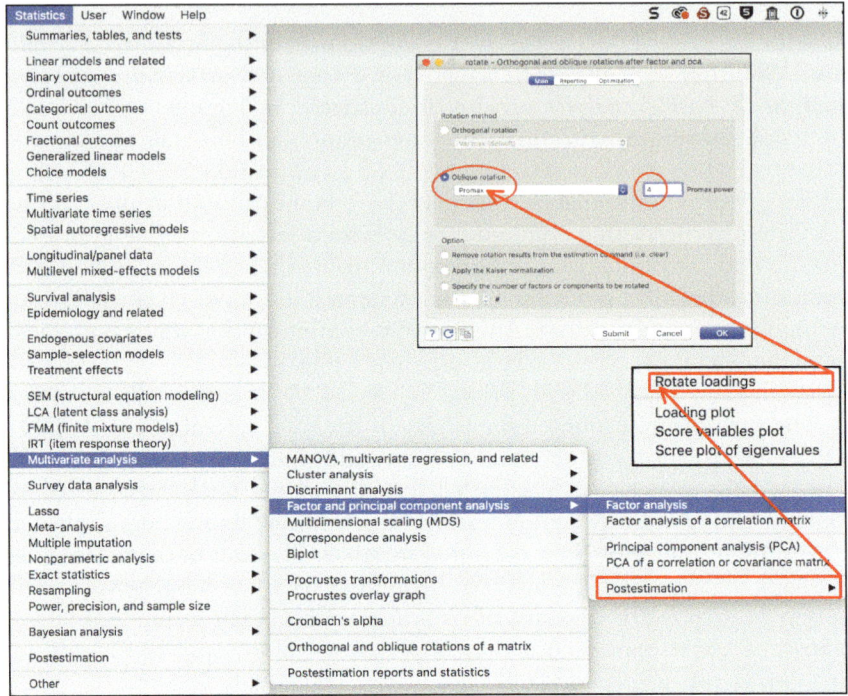

Figure 14.6 Stata factor analysis menu system

As previously noted, Stata arranges its EFA routines into two sequential sets of menus and commands. The first set, as illustrated in Figures 14.1 through 14.3, are devoted to factor extraction. Rotation and other aspects of the EFA are included in Stata's second set of menus and commands, which are called postestimation commands. Figure 14.6 displays the postestimation menus. In this case, a promax rotation was selected (Gorsuch, 2003; Tataryn et al., 1999).

Identical results are produced by the `rotate, promax` command as displayed in Figure 14.7. The pattern matrix revealed two clearly defined factors with four variables saliently loading on each factor and no complex loadings.

There are a variety of postestimation commands for which there is no menu option. For instance, the rotated and unrotated loadings can be directly compared with the `estat rotatecompare` command and the structure matrix can be displayed with the `estat structure` command. The structure coefficients presented in Figure 14.8 were strong (.56 to .92) and there was no evidence of a suppression effect (Thompson, 2004). Namely, a structure coefficient around zero but a high pattern coefficient or vice versa, or pattern and structure coefficients of different signs (Graham et al., 2003). A review of the correlation matrix revealed strong bivariate correlations between the measured variables, making it improbable that this result was simply a mathematical artifact.

```
Factor analysis/correlation                    Number of obs     =        152
   Method: maximum likelihood                  Retained factors =          2
   Rotation: oblique promax (Kaiser off)       Number of params  =         15
                                               Schwarz's BIC    =     107.59
   Log likelihood = -16.11566                  (Akaike's) AIC   =    62.2313

   ----------------------------------------------------------------------------
        Factor  |    Variance   Proportion    Rotated factors are correlated
   -------------+--------------------------------------------------------------
       Factor1  |    4.41571      0.7931
       Factor2  |    4.30201      0.7727
   ----------------------------------------------------------------------------
   LR test: independent vs. saturated:   chi2(28) =   893.09 Prob>chi2 = 0.0000
   LR test:   2 factors vs. saturated:   chi2(13) =    31.21 Prob>chi2 = 0.0031

   Rotated factor loadings (pattern matrix) and unique variances

   ------------------------------------------------------------
       Variable |  Factor1    Factor2 |   Uniqueness
   -------------+----------------------+---------------
         vocab1 |   0.9145     0.0144 |    0.1451
       designs1 |   0.1215     0.6832 |    0.4020
       similar1 |   0.7254     0.1719 |    0.2693
        matrix1 |   0.1146     0.7452 |    0.3118
        veranal2|   0.5998     0.2412 |    0.3791
         vocab2 |   0.8555     0.0443 |    0.2130
        matrix2 |   0.0956     0.7527 |    0.3234
       designs2 |   0.0198     0.7675 |    0.3892
   ------------------------------------------------------------
```

Figure 14.7 Promax rotated solution for two factors

```
. estat structure

Structure matrix: correlations between variables
and promax(3) rotated common factors

   ------------------------------------------------
       Variable |   Factor1    Factor2
   -------------+----------------------------------
         vocab1 |   0.9246     0.6559
       designs1 |   0.6008     0.7685
       similar1 |   0.8460     0.6808
        matrix1 |   0.6374     0.8256
        veranal2|   0.7690     0.6620
         vocab2 |   0.8866     0.6445
        matrix2 |   0.6236     0.8197
       designs2 |   0.5583     0.7814
   ------------------------------------------------
```

Figure 14.8 Structure matrix from two-factor solution

   Given that this was an oblique rotation, the factor correlation matrix must be ascertained. That can be accomplished with the `estat common` command, which indicated that these two factors were correlated at .70. Although elevated, this does not pose a severe threat to discriminant validity (Brown, 2015).

```
. estat factors

Factor analysis with different numbers of factors
(maximum likelihood)

----------------------------------------------------------------
#factors |    loglik     df_m    df_r       AIC        BIC
---------+------------------------------------------------------
       1 |   -60.17769      8      20    136.3554    160.5464
       2 |   -16.11566     15      13     62.23131   107.5895
       3 |    -5.689615    21       7     53.37923   116.8807
       4 |    -.5531138    26       2     53.10623   131.7271
----------------------------------------------------------------
the models with 3 4 factors are Heywood cases
```

Figure 14.9 BIC values for iq factor solutions via the estat factors postestimation command

Although the BIC (Schmitt et al., 2018) has received little research attention for use with EFA models, Stata provides it with the estat factors postestimation command for models that used ML extraction. Given that the model with the lowest BIC value is preferred, the BIC values presented in Figure 14.9 indicate that the two-factor model should be favored.

A two-dimensional factor loading plot can be requested with the loadingplot command. As illustrated in Figure 14.10, that plot vividly displays the separation between these factors.

Residual misfit is an important clue to model quality. Stata provides the residual matrix with the estat residuals command, but does not tally the number of residual coefficients greater than .05 or greater than .10 and does not compute the RMSR. With a small matrix, as in Figure 14.11, simple visual analysis may be sufficient.

However, larger matrices may be difficult to quickly scan. In that case, a standalone computer program entitled *Residuals* can be downloaded from edpsychassociates.com/Watkins3.html if detailed information about residuals is desired. The *Residuals* program requires the residual matrix as input. That can be accomplished by copying the residual matrix from the Stata results with the **Edit > Copy table** menu option and then pasting it as text into an Excel document. This spreadsheet can then be saved as a text file for input to the *Residuals* program. As shown in Figure 14.12, the RMSR value was .028 and there were no residuals >.10.

Model 2 converged properly, produced reasonable parameter estimates, and exhibited four salient loadings at good to excellent levels (Comrey & Lee, 1992) on each factor in a simple structure configuration. Communalities were high (.60 to .85; Gibson et al., 2020; Watson, 2017), RMSR was small (.028), and only two residual coefficients exceeded the absolute value of .05 (none exceeded .10). This meets the standard of close fit defined by

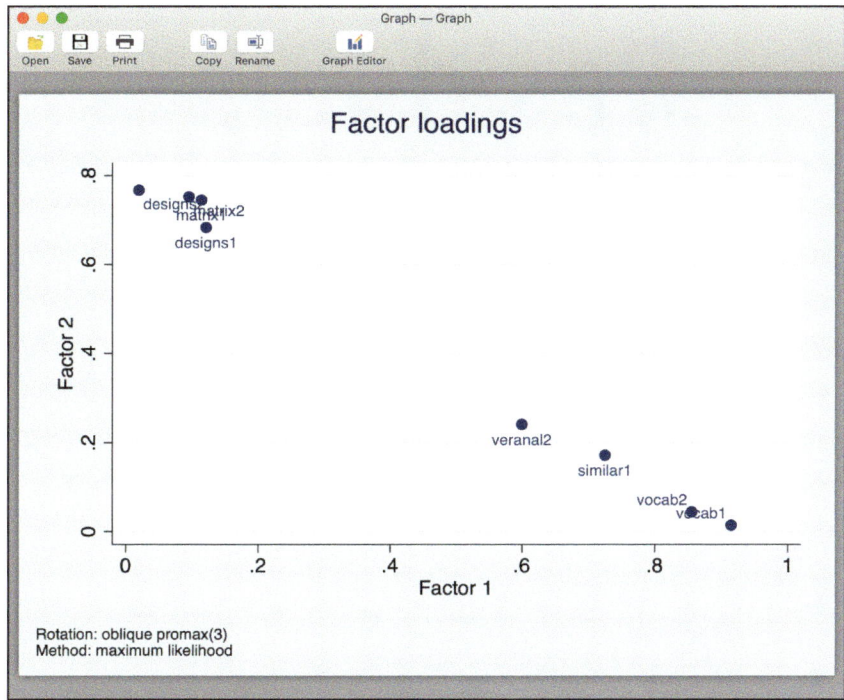

Figure 14.10 Loading plot for two–factor solution in Stata graph window

```
. estat residuals

Raw residuals of correlations (observed-fitted)

---------------------------------------------------------------------------------
     Variable |   vocab1   desig~1   simil~1   matrix1   veran~2    vocab2   matrix2   desig~2
--------------+------------------------------------------------------------------
       vocab1 |   0.0000
     designs1 |   0.0157   -0.0000
     similar1 |   0.0051    0.0018   -0.0000
      matrix1 |   0.0231    0.0085   -0.0058    0.0000
     veranal2 |  -0.0236   -0.0371    0.0299   -0.0486    0.0000
       vocab2 |   0.0041   -0.0091   -0.0186   -0.0159    0.0187   -0.0000
      matrix2 |  -0.0183   -0.0496   -0.0115    0.0323    0.0763    0.0075    0.0000
     designs2 |  -0.0144    0.0598    0.0112   -0.0273   -0.0096    0.0181   -0.0171   -0.0000
---------------------------------------------------------------------------------
```

Figure 14.11 Residual values for two–factor solution via the estat residuals postestimation command

Maydeu–Olivares (2017). The interfactor correlation of .70 was somewhat elevated but that might be due to a strong general factor that permeates all the variables (Gorsuch, 1983).

Stata can compute the alpha reliability of each set of variables identified in the EFA. Menu access is via **Statistics > Multivariate analysis > Cronbach's**

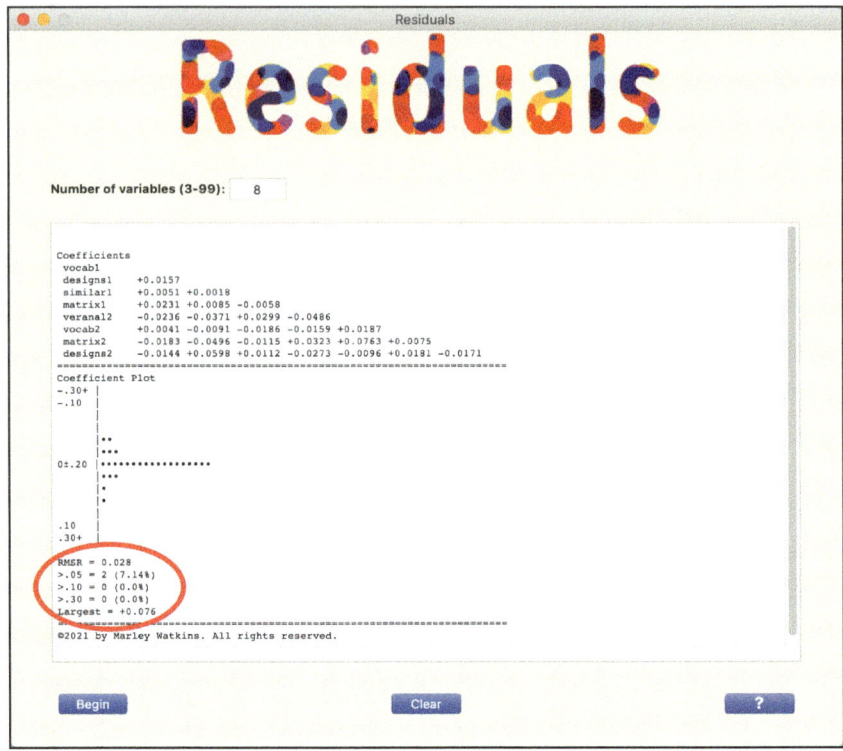

Figure 14.12 Residual matrix analysis of two-factor solution with the Residuals program

alpha and selecting the variables that comprise each factor. In this case, the alpha reliability of the first factor was computed to be .92 by the `alpha vocab1 similar1 veranal2 vocab2` command. A 95% confidence interval of for this alpha coefficient was calculated by the `cialpha` command implemented immediately following the `alpha` command. It indicated that alpha is likely greater than .90. The alpha reliability of the second factor was computed to be .88 by the `alpha designs1 matrix1 matrix2 designs2` command. Its confidence interval suggests that this alpha is likely greater than .84. Reliability coefficients of this magnitude are generally considered to be strong and might be useful for clinical decisions (DeVellis, 2017).

The two-factor model was robust to different extraction and rotation methods as very similar results were obtained when iterated principal factor extraction and oblimin rotation were employed. Additionally, the measured variables were distributed across the two factors as predicted by prior theory (verbal versus nonverbal measures) and the structure was interpretable and theoretically meaningful. In sum, the two-factor model appears to be a good EFA solution that offers the optimum balance between comprehensiveness (accounting for the most variance) and parsimony (with the fewest factors).

Model 1

Although the two-factor model appeared to be satisfactory, a one-factor model was also generated with the `factor vocab1-designs2, ml factors(1)` command to ensure that it did not exhibit superior fit characteristics. The results of that command are presented in Figure 14.13. Model 1 converged properly, produced reasonable parameter estimates, exhibited eight salient loadings on its single factor, and accounted for 61.5% of the total variance. However, uniqueness values were higher than those in Model 2.

Further, the RMSR value was .077, 57% of the residual coefficients exceeded .05, and 18% of the residuals exceeded .10. These residual values suggest that another factor might be extracted. Most critically, the single factor does not seem to reflect a common unifying theme because it encompasses both verbal and nonverbal reasoning variables. This is evidence that two factors have collapsed onto one (Fabrigar & Wegener, 2012). Thus, measures of model fit as well as theoretical convergence remove this model from consideration, leaving Model 2 as the preferred solution.

Figure 14.14 presents a do-file that summarizes this EFA.

```
Factor analysis/correlation              Number of obs     =        152
  Method: maximum likelihood             Retained factors  =          1
  Rotation: (unrotated)                  Number of params  =          8
                                         Schwarz's BIC     =    160.546
  Log likelihood = -60.17769            (Akaike's) AIC     =    136.355

      ------------------------------------------------------------------
          Factor  |  Eigenvalue  Difference        Proportion  Cumulative
      ------------+-----------------------------------------------------
         Factor1  |     4.91736           .            1.0000      1.0000
      ------------------------------------------------------------------
  LR test: independent vs. saturated:  chi2(28) =  893.09 Prob>chi2 = 0.0000
  LR test:     1 factor vs. saturated:  chi2(20) =  117.06 Prob>chi2 = 0.0000

Factor loadings (pattern matrix) and unique variances

      ---------------------------------------
         Variable |  Factor1 |   Uniqueness
      ------------+----------+---------------
           vocab1 |  0.8747  |     0.2348
         designs1 |  0.7032  |     0.5055
         similar1 |  0.8533  |     0.2718
          matrix1 |  0.7438  |     0.4468
         veranal2 |  0.7973  |     0.3644
           vocab2 |  0.8547  |     0.2695
          matrix2 |  0.7402  |     0.4521
         designs2 |  0.6799  |     0.5377
      ---------------------------------------
```

Total Variance
4.9176 ÷ 8.0 = 61.5%

Figure 14.13 Unrotated solution for one factor

```
                                    EFAIq.do
  Open   Save  Print      Find  Show  Zoom                                      Do
  1    * do-file for EFA of iq data
  2    summarize                              //descriptive statistics
  3    mvtest normality vocab1-designs2, stats(all)    //multivariate normality
  4    correlate vocab1-designs2              //Pearson correlations
  5    graph matrix vocab1-designs2           //scatterplot matrix
  6    graph box vocab1-designs2              //boxplot
  7    * Install factortest via search factortest
  8    factortest vocab1-designs2             //determinant, Bartlett, & KMO
  9    quietly pca vocab1-designs2, components(8) //set-up for postestimation
  10   screeplot                              //scree
  11   * Install fapara via ssc install fapara
  12   fapara, pca reps(500)                  //parallel analysis
  13   * Install minap via ssc install minap
  14   minap vocab1-designs2                  //MAP
  15   factor vocab1-designs2, ml factors(3) //three-factor model
  16   rotate, promax oblique                 //pattern coefficients
  17   estat structure                        //structure coefficients
  18   estat common                           //factor correlations
  19   estat residuals                        //residual matrix
  20   factor vocab1-designs2, ml factors(2) //two-factor model
  21   rotate, promax oblique                 //pattern coefficients
  22   estat structure                        //structure coefficients
  23   estat common                           //factor correlations
  24   estat residuals                        //residual matrix
  25   alpha vocab1 similar1 veranal2 vocab2   //reliability of first factor
  26   alpha designs1 matrix1 matrix2 designs2 //reliability of second factor
  27   factor vocab1-designs2, ml factors(1) //one-factor model
  28   estat residuals                        //residual matrix
  29
  Automatic    Line: 1, Col: 33
```

Figure 14.14 Stata do-file for iq EFA

Factor Names

For convenience, researchers typically name the factors identified through EFA. Names can be given to factors to facilitate the communication of results and advance cumulative knowledge. Rummel (1967) suggested that factors can be named symbolically, descriptively, or causally. Symbolic names are without any substantive meaning, for example, F1 and F2 or A, B, and C. These names merely denote the factors without adding any meaning. Descriptive names are clues to factor content that can categorize the factor in terms of its apparent content. For example, measures of word meaning, synonyms, and paragraph comprehension might be described as verbal. Causal names involve reasoning from the salient loadings to the underlying influences that caused them. Given that a factor is a construct operationalized by its factor loadings, the researcher tries to understand the underlying dimension that unifies the variables that define the factor. "Thus, the naming of a factor is based on its conceptual underpinnings" (Reio & Shuck, 2015, p. 20). For example, measures of word meaning, synonyms, and paragraph comprehension might be described as a verbal reasoning factor.

To reduce the possibility of confusion between measured variables and factors, factors should *not* be named after measured variables (Thompson, 2004). Factors are typically named by considering what their most salient measured variables have in common with higher loadings receiving greater consideration (Kahn, 2006; Thompson, 2004). Although both pattern and structure coefficients are considered, structure coefficients may be more useful for naming when the interfactor correlations are modest because they reflect the simple relationship between a variable and a factor without the confounding effect of other factors (Kahn, 2006). With high interfactor correlations, the pattern coefficients might become more useful.

Factor naming is a subjective process (Watson, 2017), and the researcher must avoid the construct identity fallacy (Larsen & Bong, 2016). That is, assuming that two factors are the same because they have the same name (jingle fallacy) or different because they have different names (jangle fallacy). In either case, the unwary might be tempted to accept an inaccurate factor label. Likewise, it would be inappropriate to reify a factor label (Cliff, 1983; Kline, 2013). That is, assume that it is a real, physical thing rather than an explanatory latent construct. "Merely because it is convenient to refer to a factor (like *g*), by use of a noun does not make it a physical thing. At the most, factors should be regarded as sources of variance, dimensions, intervening variables, or latent traits that are useful in explaining manifest phenomena, much as abstractions such as gravity, mass, distance, and force are useful in describing physical events" (Carroll, 1995b, p. 126). There is ample evidence that jingle-jangle fallacies are widespread in practice and have resulted in construct proliferation and empirically redundant measures (Le et al., 2010; Shaffer et al., 2016).

Report

Three plausible models were evaluated. It was hypothesized that the three-factor model would exhibit symptoms of overextraction. This hypothesis was confirmed by a factor with no salient loadings and an inadmissible communality estimate. In contrast, the one-factor model exhibited symptoms of underextraction: failure to reflect a common unifying theme (Fabrigar & Wegener, 2012) and many large residuals, strongly indicating the presence of another factor (Cudeck, 2000). Model 2 converged properly, produced reasonable parameter estimates, and exhibited four salient loadings on each factor in a simple structure configuration. Communalities were robust (.60 to .86) and only two off-diagonal residual coefficients exceeded the absolute value of .05. When the variables that saliently loaded each factor were combined to create a scale, their internal consistency reliability was .92 and .88.

The four measured variables that saliently loaded on the first factor contained verbal content and seemed to require reasoning with that content: defining words, describing how words are similar, and explaining verbal analogies. The four measured variables that saliently loaded on the second factor contained nonverbal content that involved analysis for patterns, inductive discovery of relationships, and deductive identification of missing components. Thus, the first factor might be called verbal reasoning and the second nonverbal reasoning. Performance on the verbal measures is obviously affected by experience, learning, and acculturation, whereas performance on the nonverbal measures might be less affected by prior learning and experience. These attributes can be found in the taxonomy of cognitive abilities elucidated by Carroll (1993) and are probably better labeled crystalized and fluid ability, respectively, to reduce construct redundancy (Le et al., 2010; Shaffer et al., 2016). The elevated interfactor correlation of .70 is also consistent with Carroll's (1993) model of a general ability that accounts for the correlation between cognitive factors (Carretta & Ree, 2001; Gorsuch, 1983).

15 Step 10

Report Exploratory Factor Analysis Results

Many published exploratory factor analysis (EFA) studies employ inappropriate methods and report inadequate methodological information (Conway & Huffcutt, 2003; Fabrigar et al., 1999: Ford et al., 1986; Gaskin & Happell, 2014; Goretzko et al., 2019; Henson & Roberts, 2006; Howard, 2016; Izquierdo et al., 2014; Lloret et al., 2017; McCroskey & Young, 1979; Norris & Lecavalier, 2010; Park et al., 2002; Roberson et al., 2014; Russell, 2002; Sakaluk & Short, 2017). EFA reports should provide a clear description of the decisions made and a comprehensive presentation of the results. In other words, the EFA process must be made transparent (Flake & Fried, 2020). Enough detail must be provided to allow "informed review, replication, and cumulation of knowledge" (Ford et al., 1986, p. 307) yet remain succinct enough for journal presentation. Ideally, "the description of the methods used in the analysis should be sufficiently clear that a reader could replicate the study exactly" (Finch, 2020a, p. 94). The Report paragraphs in prior sections are preliminary models of how each decision step could be reported. Further guidance can be obtained from author guidelines published by academic journals (e.g., Cabrera-Nguyen, 2010).

The EFA report should echo the previously enumerated decision steps. Following is an outline for the EFA report:

a. Describe the measured variables and justify their inclusion.
b. Describe the participants and justify their adequacy in terms of number and representativeness.
c. Present descriptive statistics. Ensure that linearity, outliers, and normality are addressed. Report the extent of missing data and how it was addressed. Report the correlation matrix if journal space allows. If not, present it in a supplemental table.
d. Verify that data are appropriate for EFA with Bartlett's test of sphericity, KMO sampling adequacy, and magnitude of coefficients in the correlation matrix.
e. Justify the model: principal components or common factor analysis.
f. Report the name and version of the software used to conduct the EFA. There may be differences in results due to software or version and

DOI: 10.4324/9781003149286-15

this will allow readers to understand how your results were achieved (Grieder & Steiner, 2020).

g. Detail the method of factor extraction and justify its use with the data. Report the method of estimating communalities if applicable. Report and justify the type of correlation matrix employed (Pearson, Spearman, polychoric, etc.).

h. Describe a priori criteria for determination of how many factors to retain for interpretation. Ensure that multiple criteria are applied within a model testing approach. Report the results of each criterion.

i. Identify the rotation method (orthogonal or oblique) and type (varimax, promax, oblimin, etc.). Justify those selections.

j. Interpret each model using a priori guidelines for acceptability (including salience of loadings, scale reliability, model fit standards, symptoms of over- and underextraction). For the final model, present all pattern coefficients (do not omit non-salient values), communalities, and factor intercorrelations. Also present structure coefficients if journal space allows. If not, present them in a supplemental table.

Factor Scores

It is possible to weight variable scores according to their relationship to each factor and thereby create factor score estimates for each participant that can subsequently be included in other investigations. Conceptually, a factor score is an estimate of the latent variable that underlies the measured variables. Unfortunately, "an infinite number of ways for scoring the individuals on the factors could be derived that would be consistent with the same factor loadings" (Grice, 2001, p. 431). Consequently, numerous ways to compute factor scores have been developed over the years but none have been found to be superior in all circumstances (Finch, 2013; Gorsuch, 1983; Revelle, 2016). Some methodologists prefer complex computations involving multiple regression and maximum likelihood estimation (Comrey & Lee, 1992; Hair et al., 2019). Others believe that simple unit weights (e.g., adding the scores from variables that saliently load) may be superior (Carretta & Ree, 2001; Gorsuch, 2003; Kline, 1994; Russell, 2002; Wainer, 1976). Given the current state of knowledge about factor score indeterminacy (Rigdon et al., 2019), it is best to be cautious about the use of factor scores (Osborne & Banjanovic, 2016).

Users can generate unit-weighted factor scores by simply creating a new variable using the `generate` command that is the sum of the salient variables for that factor. For example, `generate verbal = vocab1+similar1+veranal2+vocab2`. This method has been recommended when working with items within a scale (Wainer, 1976). Stata offers two more complex postestimation options to compute factor scores. Both methods use the `predict f#` command where *#* is the factor number. For example, the regression method for the two factor model can be implemented with the `predict f1 f2, regression` command, and Bartlett's (1937) method can be implemented with the `predict f1 f2, bartlett` command. Either command will generate factor scores and insert them into the **Data Editor**.

If factor scores are estimated, Nunnally and Bernstein (1994) suggested that they should exhibit strong multiple correlations with the variables that comprise each factor and should not correlate with other factor scores beyond the underlying factor intercorrelations. Mulaik (2018) recommended that factor scores be avoided unless the R^2 for predicting the common factors exceeds .95. Gorsuch (1983) recommended that the factor score correlations with factors should, at a minimum, exceed .80. In this case, the R^2 values for factors one and two were .995 and .989, respectively. An extended discussion of the benefits and liabilities of factor scores was provided by DiStefano et al. (2009).

Cautions

Factor analysis only provides models of the world. By definition, models cannot capture the complexities of the real world. "At best, they can provide an approximation of the real world that has some substantive meaning and some utility" (MacCallum, 2003, p. 115). Therefore, the researcher should not fall prey to the nominalistic fallacy or the construct identity fallacy, nor should they reify factors (Cliff, 1983; Kline, 2013; Larsen & Bong, 2016). That is, believe that naming a factor means that the factor is a real physical entity, well understood, or even correctly named. As cogently argued by Feynman (1974), the first duty of a scientist is "that you must not fool yourself—and you are the easiest person to fool" (p. 12).

A basic premise of the philosophy of science is that data do not confirm a model, they can only fail to disconfirm it (Popper, 2002). Thus, "factor analysis is not an end in itself but a prelude to programmatic research on a particular psychological construct" (Briggs & Cheek, 1986, p. 137). Factor analysis addresses only one type of construct validity evidence: the internal structural aspect that "appraises the fidelity of the scoring structure to the structure of the construct domain" (Messick, 1995, p. 745). "Strong factorial evidence is necessary, but not sufficient, for establishing evidence of validity" (McCoach et al., 2013, p. 111).

The value of factors must be judged by their replicability across samples and methods and by the meaningfulness of their relationships with external criteria (Comrey & Lee, 1992; Goldberg & Velicer, 2006; Gorsuch, 1983; Mulaik, 2018; Nunnally & Bernstein, 1994; Preacher et al., 2013). Replication is critical because results that cannot be reproduced cannot contribute to the cumulative growth of knowledge necessary for scientific progress (Open Science Collaboration, 2015). Of course, replication with independent samples would be ideal but is not always possible. If the sample is sufficiently large, it can be randomly split into two sub-samples and EFA results compared across those replication samples (Osborne & Fitzpatrick, 2012). A variety of cross-validation techniques that might be implemented with a single sample were presented by Koul et al. (2018). Absent replication, the researcher should employ alternative extraction and rotation methods to ensure that the EFA results are at least robust across those methods (Gorsuch, 1983).

It is essential that a construct validation program be implemented to delineate the external ramifications of replicated structural validity studies (Comrey & Lee, 1992; Goodwin, 1999; Gorsuch, 1983; Lubinski & Dawis, 1992; Rencher & Christensen, 2012). See Benson (1998) and Simms and Watson (2007) for tutorials on construct validation programs and Messick (1995) for a discussion of construct validity.

16 Exploratory Factor Analysis with Categorical Variables

The measured variables in the original application of exploratory factor analysis (EFA) by Spearman (1904) were scores on school tests of math, spelling, etc. Thus, each variable was the sum of multiple math or spelling items called a scale and EFA was developed for the analysis of such scales (Gorsuch, 1997). Scales are more reliable than items because they rely on the aggregation principle whereby common variance accumulates whereas measurement error, being random, does not (Lubinski & Dawis, 1992). Additionally, response options for items often take the form of a set of ordered categories rather than a continuous range of values. For ability and achievement items, responses may allow only two categories: correct or incorrect. For attitude items, 3- to 7-category Likert-type responses (*strongly agree* to *strongly disagree*) are often used with five alternatives the most common (Likert, 1932). These ordered categorical responses constitute ordinal variables (Stevens, 1946). The psychometric characteristics of items (compared to scales) require careful consideration when conducting an EFA (Gorsuch, 1997; Nunnally & Bernstein, 1994; Reise et al., 2000; Widaman, 2012).

Data

The sdq.xlsx file can be imported via *File > Import > Excel spreadsheet* menu options, the imported data can be viewed with the browse command, and the properties of the sdq variables can be observed with the describe and codebook commands. These commands reveal that the 30 sdq items are numeric with values of one through six and there are no missing data indicators. As previously described, these items were measured at the ordinal level.

Participants

Participants were 425 elementary school children. The communality of the variables under study is unknown, but prior international research and the ordinal nature of the response scale indicate that it would be reasonable to estimate communality to be in the low range (Mucherah & Finch, 2010). With 30 items and three anticipated factors, the number of items per factor

DOI: 10.4324/9781003149286-16

is 10. Rouquette and Falissard (2011) simulated typical attitudinal scale data and reported that scales with that item to factor ratio required a sample of 350 to 400 participants. The ratio of participants to measured variables is 14 to 1, exceeding the recommendations of several measurement specialists (Child, 2006; Gorsuch, 1983). Given these considerations, the current sample size of 425 participants was judged to be adequate.

Data Screening

Anything that influences the correlation matrix can potentially affect EFA results (Carroll, 1985; Flora et al., 2012; Onwuegbuzie & Daniel, 2002). Accordingly, the data must be screened before conducting an EFA to ensure that some untoward influence will not bias EFA results (Flora et al., 2012; Goodwin & Leech, 2006; Hair et al., 2019; Walsh, 1996).

Pearson versus Polychoric Correlations

Pearson correlations assume bivariate normality, which requires continuous scores (Sheskin, 2011). Thus, categorical scores are not, by definition, normally distributed (Bandalos, 2018; Hancock & Liu, 2012; Puth et al., 2015; Walsh, 1996). In addition, categorization of continuous scores causes imprecise estimation of Pearson correlations. For example, Bollen and Barb (1981) used simulated data and demonstrated the effect of categorizing two continuous scores with a correlation of .90. As illustrated in Figure 16.1, the reduction in precision is relatively modest for seven categories, falling from .90 to .85. However, the fewer the categories, the more imprecise the correlation estimates. With only two categories, the estimated correlation dropped to .72, a reduction of 20%.

Additionally, the Pearsonian correlation for dichotomous items (phi) is strongly influenced by differences in item endorsement frequencies. Rummel (1970) suggested deleting dichotomous variables with > 90 to 10 splits between categories because the correlation coefficients between these variables and others are truncated and because the scores for the cases in the small category are more influential than those in the category with numerous cases. Of course, imprecise correlation estimates impact EFA results.

Given that EFA is conducted on a correlation matrix, it is important to utilize the optimal type of correlation for ordinal data. Bollen and Barb (1981) suggested "that under certain conditions it may be justifiable to use Pearson correlations and analyze categorical data as if it were continuous" (p. 232). Lozano et al. (2008) conducted a statistical simulation study and found that four response categories were minimal and seven optimal to ensure adequate factorial validity. Other methodologists have suggested the use of Pearson correlations when there are at least five ordered categories (DiStefano, 2002; Mueller & Hancock, 2019).

However, nonnormality can be problematic for Pearson correlation estimates (Flora et al., 2012; Pett et al., 2003; Sheskin, 2011; Yong & Pearce,

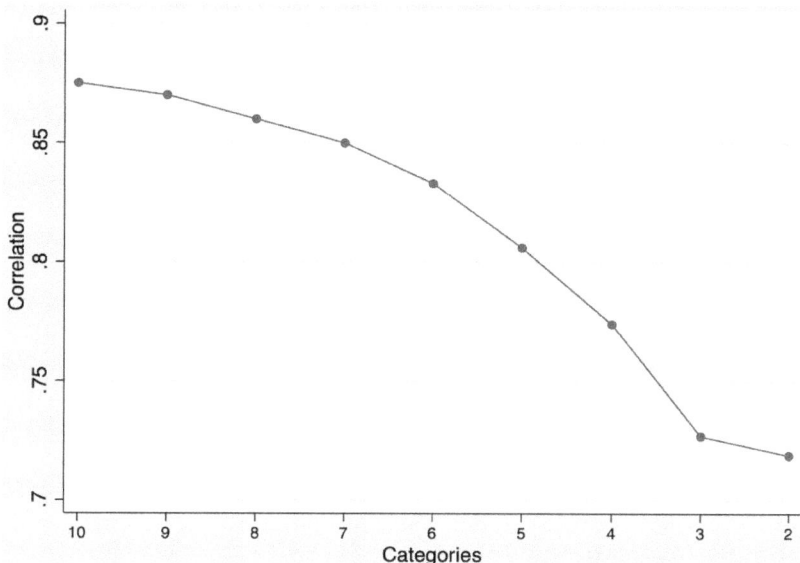

Figure 16.1 Effect on correlation coefficient of categorizing continuous variables

2013). The extent to which variables can be nonnormal and not substantially affect EFA results has been addressed by several researchers. Curran et al. (1996) opined that univariate skew should not exceed 2.0 and univariate kurtosis should not exceed 7.0. Other measurement specialists have agreed with those guidelines (Bandalos, 2018; Fabrigar et al., 1999; Wegener & Fabrigar, 2000). In terms of multinormality, statistically significant multi-variate kurtosis values > 3.0 to 5.0 might bias factor analysis results (Bentler, 2005; Finney & DiStefano, 2013; Mueller & Hancock, 2019). Thus, Pearson correlations might not be appropriate for ordinal data with five to seven cat-egories if the variables are severely nonnormal.

In fact, some methodologists assert that "it is often not appropriate to pre-tend that categorical variables are continuous" (Flora et al., 2012, p. 12) and have studied the characteristics of alternative types of correlation coefficients for factor analysis of ordinal data. These alternatives include the non-parametric correlations of Kendall and Spearman (Revelle, 2016) as well as polychoric and tetrachoric correlations (Basto & Pereira, 2012; Choi et al., 2010; McCoach et al., 2013).

Polychoric correlations assume that a normally distributed continuous, but unobservable, latent variable underlies the observed ordinal variable and are a maximum likelihood estimate of the Pearson correlations for those underlying normally distributed continuous variables (Choi et al., 2010). A tetrachoric correlation is a special case of the polychoric correlation applicable when the observed variables are dichotomous (Panter et al., 1997). Although some researchers prefer Spearman correlations for kurtotic distributions or when outliers are present (de Winter et al., 2016), factor analysis results based on polychoric correlations have better reproduced the measurement model than Pearsonian correlations (Barendse et al., 2015; Carroll, 1961; Choi et al., 2010; Flora et al., 2012; Flora & Flake, 2017; Holgado-Tello et al., 2010 Lloret et al., 2017; van der Eijk & Rose, 2015; Zhang & Browne, 2006). In fact, Flora and Curran (2004) found that polychoric correlations tended to be robust to violations of univariate nonnormality until nonnormality became extremely severe (skew = 5.0, kurtosis = 50.0). However, polychoric correlations are estimates of the latent correlation between assumed continuous variables and that estimation process can sometimes go amiss, producing imprecise estimates or a statistically improper matrix (Lorenzo-Seva & Ferrando, 2020). If there are few or no responses to more than 20% of the items on a factor then it may be beneficial to recode the data into fewer categories for estimation of the polychoric matrix (DiStefano et al., 2021). Further, polychoric correlations would not be appropriate if the assumption of an underlying normally distributed latent variable is not tenable.

Given the ordinal nature of the sdq data, it was anticipated that nonnormality would be readily apparent. Screening of the sdq data begins with a review of the univariate descriptive statistics via the `tabstat sdq1-sdq30, statistics(count mean sd min max skewness kurtosis)` command.

The resulting descriptive statistics are displayed in Figure 16.2. The minimum and maximum values all range from one to six, indicating no illegal or out-of-bounds values. There is no missing data (the *n* for each item is 425) and there was no item with zero variance. The mean values are high (many in the four to five range for a scale with a maximum score of six).

stats	sdq1	sdq2	sdq3	sdq4r	sdq5r	sdq6	sdq7	sdq8	sdq9r	sdq10r
N	425	425	425	425	425	425	425	425	425	425
mean	4.169412	4.832941	4.816471	3.818824	5.317647	4.004706	4.054118	4.769412	4.856471	4.792941
sd	1.796093	1.371147	1.422266	1.92001	1.27023	1.814506	1.798686	1.38336	1.578336	1.580974
min	1	1	1	1	1	1	1	1	1	1
max	6	6	6	6	6	6	6	6	6	6
skewness	-.5677836	-1.208667	-.9644799	-.2458382	-1.952322	-.448097	-.4827749	-1.21438	-1.272523	-1.162776
kurtosis	1.951858	3.72238	2.816773	1.513894	5.874939	1.832158	1.867043	3.812607	3.389532	3.136582

stats	sdq11r	sdq12	sdq13	sdq14	sdq15r	sdq16r	sdq17r	sdq18	sdq19	sdq20
N	425	425	425	425	425	425	425	425	425	425
mean	5.023529	4.437647	3.847059	4.774118	5.143529	4.296471	4.962353	4.190588	4.235294	4.807059
sd	1.293821	1.484668	1.866448	1.382433	1.457127	1.795332	1.594062	1.788038	1.693822	1.452209
min	1	1	1	1	1	1	1	1	1	1
max	6	6	6	6	6	6	6	6	6	6
skewness	-1.391265	-.8374838	-.3107814	-1.279837	-1.660699	-.6718852	-1.445885	-.5753974	-.624968	-1.267053
kurtosis	4.282132	2.786191	1.655025	4.005897	4.496659	2.04355	3.773838	1.965232	2.12792	3.740691

stats	sdq21r	sdq22r	sdq23r	sdq24	sdq25	sdq26	sdq27r	sdq28r	sdq29r	sdq30
N	425	425	425	425	425	425	425	425	425	425
mean	5.042353	4.868235	5.131765	4.463529	3.967059	5.517647	3.818824	4.684706	5.576471	4.710588
sd	1.532077	1.663014	1.467109	1.549258	1.777439	1.011861	1.881545	1.757565	.9612353	1.454825
min	1	1	1	1	1	1	1	1	1	1
max	6	6	6	6	6	6	6	6	6	6
skewness	-1.524956	-1.339722	-1.668782	-.8766241	-.4674258	-2.769484	.1753328	-1.032422	-2.594656	-1.113123
kurtosis	4.07741	3.414229	4.603256	2.771164	1.852379	11.19345	1.51135	2.597595	9.302301	3.374671

Figure 16.2 Descriptive statistics for sdq data

Univariate skew reflects that distributional tilt. For Stata, normal skew is zero and normal kurtosis is three. Thus, two variables exhibited high skew (–2.77 and –2.59) and high kurtosis (11.19 and 9.30), reflective of severe univariate nonnormality (Curran et al., 1996).

Multivariate normality can be ascertained with the `mvtest normality sdq1-sdq30, stats(all)` command. The expected value of Mardia's kurtosis for a multivariate normal population is $v(v + 2)$ where v is the number of variables. Thus, the expected kurtosis is 960 and Mardia's (1970) multivariate kurtosis was 1,208.1 ($\chi^2[1] = 3{,}406.2$, $p < .001$). As predicted by the univariate distributions, the data are not multivariate normal.

A Pearson correlation matrix might be employed because the data have six ordered categories which should make them continuous-enough (DiStefano, 2002; Lozano et al., 2008; Mueller & Hancock, 2019). Using the `corrci` commands as listed in the Figure 9.1 do-file, the minimum, maximum, and mean Pearson coefficients were –.16, .69, and .18, respectively.

However, the violation of multivariate normality indicates that a polychoric correlation matrix might be more appropriate input for this data. Stata provides routines for Pearson, Spearman, Kendall, and tetrachoric correlations but an ado-file must be located and downloaded to produce polychoric correlations. That file can be located with the `search polychoric` command, downloaded by selecting the ***polychoric from http://staskolenikov.net/stata*** link, and implemented with the `polychoric sdq1-sdq30` command.

The Pearson correlation between the sdq1 and sdq2 variables was .134 and the Pearson correlation between sdq1 and sdq4r was .575. The first portion of the polychoric matrix is displayed in Figure 16.3, showing that those same polychoric coefficients were .138 and .631, respectively.

Polychoric correlation matrix

	sdq1	sdq2	sdq3	sdq4r	sdq5r	sdq6	sdq7	sdq8
sdq1	1							
sdq2	.13764623	1						
sdq3	-.05765067	.26392284	1					
sdq4r	.63054347	.0781462	.02667618	1				
sdq5r	.19016314	.43574204	.26627421	.24152361	1			
sdq6	-.01362491	.14386597	.39558368	-.10783655	.06034531	1		
sdq7	.63213001	.1377119	.00189454	.40144686	.16128071	.25033448	1	
sdq8	.03689392	.48200013	.31960688	.07998888	.34524998	.23189673	.09846518	1
sdq9r	.01189908	.26490515	.43226555	.11941159	.30890118	.17596249	.04155711	.17305205
sdq10r	.56225567	.16105665	.18852126	.61265359	.28294493	-.09693782	.32875951	.13877194
sdq11r	.14126789	.45368434	.39977853	.21214931	.52289548	.20902048	.15559839	.38041946
sdq12	-.11085475	.21464211	.41233969	-.09523677	.08224259	.37161702	-.03433227	.27572545
sdq13	.7544683	.1784015	.0121302	.50386232	.23514642	.11885621	.69993019	.14836338
sdq14	.11601512	.49355516	.33564273	.09774459	.33461154	.14246671	.12708772	.61937784
sdq15r	-.01593997	.23467373	.39873181	.08272095	.30513125	.04710705	-.05506678	.2159784
sdq16r	.62794074	.0980729	-.00037645	.67564011	.14468232	-.11256904	.42811097	.04412359
sdq17r	.13886293	.63402514	.24382023	.07128993	.46187511	.06411668	.11292063	.37547159
sdq18	-.11503229	.14958857	.52405648	-.17639148	.07030139	.58667348	-.01638252	.26670777
sdq19	.69340985	.06988297	.00015848	.61367218	.21294327	-.06874379	.48913561	.02618309
sdq20	.15330392	.30741762	.25959783	.0651683	.2908362	.0574294	.1744633	.40044613
sdq21r	.00154677	.14802538	.37471977	-.03900942	.18454924	.30487658	.07457884	.23613949
sdq22r	.54738996	.14185595	.1562694	.42502566	.16160972	.09017729	.57099385	.19557611
sdq23r	.22496945	.48518989	.19204226	.12775678	.50000123	.16785256	.15194748	.3702295
sdq24	-.1012781	.16381335	.57087342	-.07518131	.14975224	.37722649	-.05480125	.24368416
sdq25	.66374444	.09661807	-.05954395	.53327966	.2258535	-.01600863	.45077854	.05454902
sdq26	.14488326	.27620526	.23683939	.08149531	.23566836	.01671492	.17582925	.41471795
sdq27r	.06345853	.28166912	.31912945	.04474539	.23355066	.17229981	-.08659581	.27397701
sdq28r	.72097592	.1724153	.04084808	.54433131	.24785093	.00288434	.60811931	.10768614
sdq29r	.23339596	.52502639	.27878098	.16362358	.64722887	.12368923	.17344148	.42059412
sdq30	-.14259167	.25734929	.51781899	-.09971429	.09103273	.40584772	-.0414595	.31743339

Figure 16.3 Polychoric correlation coefficients for sdq data

Given that these are ordinal data, it might be useful to view frequency distributions or cross-tabulations of responses (McCoach et al., 2013). Although each item has six response options, it is possible that the participants never selected one or more options, effectively reducing the actual number of response options (empirical underidentification).

Frequency tables for each variable can be produced via **Statistics > Summaries, tables, and tests > Frequency tables > Multiple one-way tables** menu options. Commands are easier. For example, `tab1 sdq1-sdq30`. The Frequencies output is quite lengthy so only the first two frequency tables are displayed in Figure 16.4.

A review of all 30 tables reveals that items 26 and 29 had few responses for some categories, but both had at least one participant for all six options. These are the same two variables that exhibited severe skew and kurtosis (see Figure 16.2). If desired, more intelligible tables can be obtained from the crosstabs procedure via **Statistics > Summaries, tables, and tests > Frequency tables > Two-way table with measures of association** menu options or the `tabulate sdq1 sdq2` command.

```
-> tabulation of sdq1

    sdq1 |      Freq.       Percent         Cum.
---------+-----------------------------------------
       1 |         62         14.59        14.59
       2 |         25          5.88        20.47
       3 |         62         14.59        35.06
       4 |         51         12.00        47.06
       5 |         80         18.82        65.88
       6 |        145         34.12       100.00
---------+-----------------------------------------
   Total |        425        100.00

-> tabulation of sdq2

    sdq2 |      Freq.       Percent         Cum.
---------+-----------------------------------------
       1 |         17          4.00         4.00
       2 |         18          4.24         8.24
       3 |         33          7.76        16.00
       4 |         63         14.82        30.82
       5 |        114         26.82        57.65
       6 |        180         42.35       100.00
---------+-----------------------------------------
   Total |        425        100.00
```

Figure 16.4 Frequency tables for sdq data

Outliers

Ordinal scale data is often obtained from surveys that rely on respondents to provide honest and thoughtful answers. Thus, it is assumed that respondents were motivated to respond accurately. Unfortunately, that assumption may not be correct and participants may provide invalid responses due to linguistic incompetence, deliberate misrepresentation, or careless inattentiveness (Curran, 2016). Linguistic incompetence is related to how the scale was constructed and its proper application to a specific sample. For example, asking preschool children to respond to a written self-concept scale would likely produce invalid responses because most preschool children cannot read. Misrepresentation typically involves respondents cheating or faking (either good or bad), usually on high-stakes tests or surveys (Curran, 2016), whereas carelessness is the result of responding without consideration of the content of the items (Dunn et al., 2018).

A variety of screening techniques have been developed to detect invalid responses (Curran, 2016; DeSimone et al., 2015; Dunn et al., 2018), but only the Mahalanobis distance (D^2) as a general method of outlier detection is available with Stata. After installing the ado-file with the `ssc install mahapick` command, the `mahascore sdq1-sdq30, gen(mahd) refmeans compute_invcovarmat` command will generate a new variable named mahd that contains the D^2 value for each case. An id variable can be created with the `gen case_id = _n` command, followed by sorting the D^2 values from low to high with the `sort mahd` command. The `browse` command allows that variable to be viewed as in Figure 16.5.

D^2 follows a chi-square distribution with degrees of freedom equal to the number of variables and can, therefore, be tested for statistical significance but "it is suggested that conservative levels of significance (e.g., .005 or .001) be used as the threshold value for designation as an outlier" (Hair et al., 2019, p. 89). Stata can compute the critical value of a chi-square distribution. Thus, the critical value for probability of .001 and 30 degrees of freedom is computed as 59.7 by the `display invchi2tail(30,.001)` command.

Figure 16.5 revealed that case #187 has the largest D^2 value (96.89), case #416 the next largest (84.68), etc. Using the conservative critical level of $p <$ .001, there are more than 20 cases that might be outliers. There is no obvious explanation for why these values are discrepant. As recommended by Hair et al. (2019), data "should be retained unless demonstrable proof indicates that they are truly aberrant and not representative of any observations in the population" (p. 91). However, as with continuous data (Bandalos & Finney, 2019; Leys et al., 2018; Tabachnick & Fidell, 2019; Thompson, 2004), results with and without the outlier data should be reported (Curran, 2016; DeSimone et al., 2015; Dunn et al., 2018). Inconsistent results may raise questions about the scale or the sample whereas consistent results will allow greater confidence in the results. In this case, the full data set is retained for subsequent analyses.

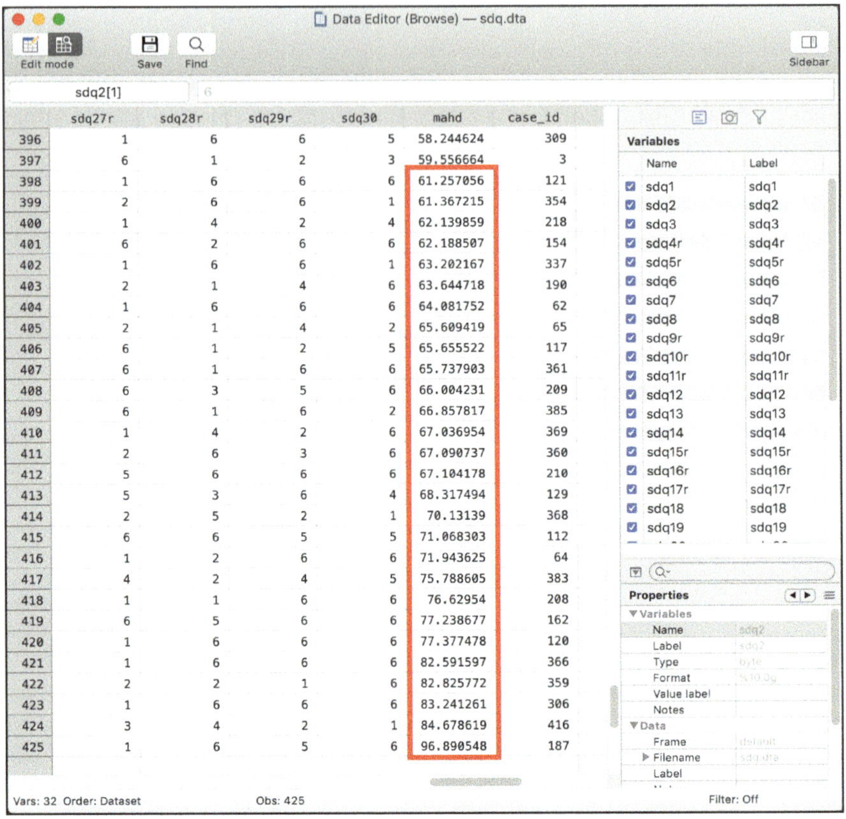

Figure 16.5 Data editor window with case id and Mahalanobis distance variables

Is EFA Appropriate?

Correlation coefficients, the determinant, Bartlett's test of sphericity (1950), and the Kaiser-Meyer-Olkin measure of sampling adequacy (KMO; Kaiser, 1974) will be considered. A quick visual scan of the polychoric matrix (Figure 16.3) found many polychoric coefficients ≥ .30 (Hair et al., 2019; Tabachnick & Fidell, 2019).

The appropriateness of a correlation matrix for EFA can be evaluated with a user-contributed ado-file named factortest that can be installed via `ssc install factortest` and implemented with the `factortest sdq1-sdq30` command. As displayed in Figure 16.6, the determinant is reported to be .000 but this figure is specious. Stata only reported these results to three decimal places so the determinant is less than .001. Bartlett's test of sphericity (1950) also tests whether the determinant of the matrix is

```
Determinant of the correlation matrix
Det                  =      0.000


Bartlett test of sphericity

Chi-square           =              5346.271
Degrees of freedom =                   435
p-value              =                 0.000
H0: variables are not intercorrelated


Kaiser-Meyer-Olkin Measure of Sampling Adequacy
KMO                  =      0.884
```

Figure 16.6 Determinant, Bartlett test of sphericity, and KMO for sdq data

zero. In this case, it statistically rejected the hypothesis that the correlation matrix was an identity matrix (chi-square of 5346.3 with 435 degrees of freedom) at $p < .001$ (if a probably value is smaller than the specified decimal precision then Stata displays a probability value of .000 but it is NOT zero). Therefore, the determinant was not zero. Additionally, the KMO measure of sampling adequacy (Kaiser, 1974) was acceptable at .88 Altogether, these measures indicated that the Pearson correlation matrix is appropriate for EFA (Hair et al., 2019; Tabachnick & Fidell, 2019). It is assumed that the polychoric matrix is similarly appropriate.

Factor Analysis Model

The purpose of this study was to uncover the latent structure underlying the 30 SDQ items. Accordingly, a common factor model (EFA) was selected (Widaman, 2018).

Factor Extraction Method

Maximum likelihood estimation can be biased in the presence of multi-variate nonnormality. Therefore, an estimation method with greater computational robustness and reduced sensitivity to nonnormality would be preferred (Barendse et al., 2015; Cudeck, 2000; Lee et al., 2012; Rhemtulla et al., 2012; Zhang & Browne, 2006). Accordingly, iterated principal factor extraction was conducted with initial communalities estimated by squared multiple correlations (Tabachnick & Fidell, 2019), which are the default in Stata.

How Many Factors to Retain

Empirical guidelines include parallel analysis (PA), minimal average partials (MAP), and scree (Velicer et al., 2000). Parallel analysis with Pearson correlations has been shown to perform well with ordinal data (Cho et al., 2009; Garrido et al., 2013). Stata considers PA, MAP, and scree to be postestimation methods so an initial PCA or EFA must be conducted to allow computation of these empirical guidelines. A series of three commands will be used for that initial analysis: (a) `polychoric sdq1-sdq30`; (b) `matrix C = r(R)`; and (c) `factormat C, n(425) factors(6) ipf`. Note that the command for factor analysis of correlation matrices is `factormat` whereas the command for raw data is `factor`. The postestimation `estat kmo` command revealed that the overall KMO value for the polychoric matrix was .87, almost identical to the KMO from the Pearson matrix.

As previously detailed, PA and MAP are available through user-contributed ado-files called fapara and minap, respective, that can be installed with the `ssc install fapara` and `ssc install minap` commands. Parallel analysis can then be performed with the `fapara, pca reps(500)` command. When the eigenvalues from random data were compared to the eigenvalues from the sdq data (Figure 16.7), the fourth random eigenvalue (1.35) was larger than the fourth real eigenvalue (1.22). Thus, parallel analysis indicated that three factors should be sufficient.

PA can also be conducted with online calculators available at www. statstodo.com/ParallelAnalysis_Exp.php and https://analytics.gonzaga. edu/parallelengine. Alternatively, a standalone computer program called

```
PA -- Parallel Analysis for Principal Components -- N = 425
PA Eigenvalues Averaged Over 500 Replications
            PCA           PA            Dif
  1.     7.670743  >  1.527458        6.143285
  2.     5.246713  >   1.45398        3.792732
  3.     2.130487  >  1.398706         .7317812
  4.      1.2187   <  1.352525        -.1338251
  5.      .7729037    1.310976        -.5380723
  6.      .6099587    1.272571        -.6626124
  7.      .4301149    1.23516         -.8050447
  8.      .3580465    1.202123        -.8440766
  9.      .2781351    1.168519        -.8903843
 10.      .2207104    1.136996        -.9162853
::::::::::::::::::::::::::::::::::::::::::
 30.     -.2873344    .5733258        -.8606603
```

Figure 16.7 Parallel analysis results from Stata fapara ado-file

Monte Carlo PCA for Parallel Analysis can be downloaded from http://edpsychassociates.com/Watkins3.html.

MAP can be implemented with the `minap, corr(C)` command. As displayed in Figure 16.8, the MAP value sequentially decreased from .0691 for one component, to .0298 for two components, to .0191 for three components, to .0175 for four components, and then increased to .0178 for five components. The lowest MAP value identifies the number of factors to retain. In this case, MAP reached a minimum at four components. Thus, MAP suggests extraction of one more factor than PA.

The scree plot can be displayed with the `screeplot` command. As often happens, the scree plot is ambiguous (Mulaik, 2018; Streiner, 1998). Figure 16.9 shows why the scree plot is considered subjective: analysts might reasonably posit different numbers of factors to extract. In this case, perhaps as many as six factors or as few as three factors.

The sdq items are thought to measure three aspects of self-concept. In theory, then, there should be three factors. However, PA suggested three factors, MAP indicated that four factors would be sufficient, and the scree plot signaled three to six factors. Therefore, models with six (the largest estimate, obtained from the visual scree), five, four, three, and two factors will be sequentially evaluated for their interpretability and theoretical meaningfulness (Preacher et al., 2013).

```
Minimum Average Partial Correlation

   m = 0        f0 =          .09587225

   m = 1        f1 =          .06906377
   m = 2        f2 =          .02983760
   m = 3        f3 =          .01907581
   m = 4        f4 =          .01752461
   m = 5        f5 =          .01781784
   m = 6        f6 =          .01857435
   m = 7        f7 =          .01990393|
   m = 8        f8 =          .02167075
   m = 9        f9 =          .02460092
   m = 10       f10 =         .02747300
   ::::::::::::::::::::::::::::::::::
   m = 29       f29 =         1.0000000

minap procedure suggests that 4
principal components should be extracted
```

Figure 16.8 Minimum average partial results from Stata minap ado-file

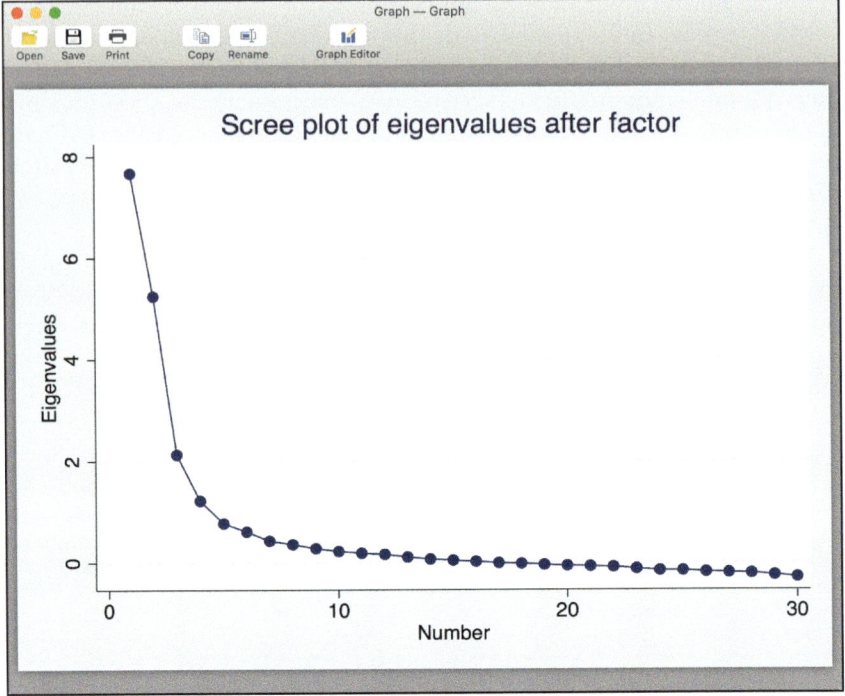

Figure 16.9 Scree plot of sdq data

Rotate Factors

An oblique rotation was selected because it honors the ubiquity of intercorrelations among social science variables (Meehl, 1990). Among the potential oblique analytic rotations, "promax rotation is almost always a good choice" (Thompson, 2004, p. 43).

Interpret Results

Given the oblique rotation, pattern coefficients and factor intercorrelations will receive primary attention during the model evaluation process (Bandalos, 2018; Hair et al., 2019). To ensure both practical (10% variance) and statistical ($p < .01$) significance of pattern loadings, the threshold for salience will be set at .32 (Norman & Streiner, 2014) with a goal of approximate simple structure (Morin et al., 2020; Thurstone, 1947). This threshold seems reasonable given that a meta-analysis of EFA outcomes found that the average factor loading of ordinal data was around .32 (Peterson, 2000).

Each model will be examined for symptoms of overextraction such as improper solutions, fewer than three salient loadings, technical factors, etc.

```
                              sdqEFA.do
 Open   Save  Print        Find  Show  Zoom                                    Do
  1    * SDQ EFA commands
  2    * download and install factortest ado-file
  3    * download and install fapara ado-file
  4    * download and install minap ado-file
  5    * download and install polychoric ado-file
  6    factortest sdq1-sdq30   //Determinant, Bartlett's test, & KMO
  7    polychoric sdq1-sdq30   //Compute polychoric correlation matrix
  8    matrix C = r(R)         //use polychoric matrix for later EFAs
  9    * EFA with 6 factors using polychoric matrix
 10    factormat C, n(425) factors(6) ipf
 11    * Postestimation commands for this model
 12    fapara, pca reps(500)   //Parallel analysis with 500 reps
 13    minap, corr(C)          //MAP
 14    screeplot               //scree
 15    rotate, promax oblique  //promax oblique rotation
 16    estat common            //factor correlations
 17    estat structure         //structure matrix
 18    estat kmo               //sampling adequacy for all variables
 19    estat residuals         //residual matrix
 20    * Change number of factors in line 10 for next model
 21    factormat C, n(425) factors(5) ipf
 22    *Repeat lines 15-19 for each model
 Automatic    Line: 1, Col: 67
```

Figure 16.10 Stata do-file for sdq EFA

as well as for symptoms of underextraction such as no common unifying theme, many complex loadings, etc. (Bandalos, 2018; Fabrigar & Wegener, 2012). Finally, the alpha reliability of scales created from the salient variables of each factor should reach .80 given that the intended use of these variables is for group research (DeVellis, 2017).

The use of commands can make this series of model comparisons relatively easy. The commands in Figure 16.10 can be edited for each model. Thus, line 10 can be changed to "factors(5)" for the five-factor model, to "factors(4)" for the four-factor model, etc.

Item content may be a useful reference when interpreting factor models. The sdq items are color-coded in Figure 16.11 to assist in that task.

The six-factor model converged properly, produced reasonable parameter estimates, and accounted for 58.8% of the total variance before rotation as computed in Figure 16.12.

However, the six-factor model exhibited symptoms of overextraction after rotation (Figure 16.13). The first factor exhibited ten salient loadings in agreement with a math self-concept scale, the second factor exhibited eight salient loadings congruent with a general self-concept scale, and the third factor exhibited seven salient loadings consistent with a verbal self-concept scale. The fourth factor contained three items that specifically referenced reading that were originally intended to load on the verbal factor. The final two factors in the six-factor model were saliently loaded by three or four items but most were complex with salient loadings on other factors. A similar pattern was produced by the five-factor model. Given these weak factors, the five- and six-factor models were not considered plausible.

No	Description	No	Description
1.	Math is my best subject	16.	Do badly on math tests
2.	Overall, I'm proud	17.	Not much to be proud of
3.	Hopeless in English class	18.	English is one of best subjects
4.	Need help in math	19.	Good grades in math
5.	Overall, I'm no good	20.	Do things as well as most
6.	Look forward to English class	21.	I hate reading
7.	Look forward to math class	22.	Never want another math course
8.	Most things I do well	23.	My life is not very useful
9.	Do badly on reading tests	24.	Good grades in English
10.	Trouble understanding math	25.	Always done well in math
11.	Nothing ever turns out right	26.	Can do almost anything if try
12.	English class is easy	27.	Trouble with writing
13.	I enjoy studying math	28.	Hate math
14.	Most things turn out well	29.	Overall I'm a failure
15.	Not good at reading	30.	Learn quickly in English class

Figure 16.11 Color-coded sdq items

```
Factor analysis/correlation                Number of obs   =       425
Method: iterated principal factors         Retained factors =        6
Rotation: (unrotated)                      Number of params =      165

                                                   Common variance
    ----------------------------------------    ---------------------------
       Factor  |  Eigenvalue  Difference  Total  Proportion   Cumulative
    -----------+----------------------------    ---------------------------
       Factor1 |    7.67074     2.42403   25.6%    0.4346      0.4346
       Factor2 |    5.24671     3.11623   17.5%    0.2973      0.7319
       Factor3 |    2.13049     0.91179    7.1%    0.1207      0.8526
       Factor4 |    1.21870     0.44580    4.1%    0.0691      0.9217
       Factor5 |    0.77290     0.16295    2.6%    0.0438      0.9654
       Factor6 |    0.60996     0.17984    2.0%    0.0346      1.0000
    Total variance = 17.6495 ÷ 30 = 58.8%
    ---------------------------------------------------------------------
```

Figure 16.12 Unrotated solution for six factors

The first three factors in the four-factor model were consistent with the theoretical structure of the scale (Figure 16.14). The items containing math and general self-concept content consistently cohered across these three models whereas the items designed to tap verbal self-concept tended to split into reading versus English content groupings. The final factor in the four-factor model was saliently loaded by only three items: one was complex and

Rotated factor loadings (pattern matrix) and unique variances

Variable	Factor1	Factor2	Factor3	Factor4	Factor5	Factor6	Uniqueness
sdq1	0.8926	0.0349	0.0385	-0.1534	-0.0698	0.1088	0.2366
sdq2	-0.0669	0.6954	0.0421	-0.0597	0.0884	-0.0045	0.4914
sdq3	-0.0142	0.0958	0.4769	0.2709	0.0724	-0.0173	0.4913
sdq4r	0.7546	-0.0318	0.0026	-0.0223	-0.0653	-0.1906	0.4065
sdq5r	0.0578	0.7008	-0.0521	0.0976	-0.1197	-0.0722	0.4857
sdq6	0.0556	0.1140	0.5993	-0.0053	-0.1960	0.4378	0.4689
sdq7	0.6799	0.0169	0.0471	-0.0512	0.0165	0.4816	0.3524
sdq8	-0.0765	0.3820	0.1436	-0.0774	0.4379	0.0319	0.4916
sdq9r	0.0336	0.1073	0.1760	0.5629	-0.1289	-0.0435	0.5296
sdq10r	0.6361	0.0887	-0.0582	0.0895	0.0595	-0.2516	0.3912
sdq11r	0.0842	0.6100	0.0216	0.1378	0.0257	0.0436	0.4780
sdq12	-0.0155	-0.0485	0.7380	-0.0331	0.0832	0.0227	0.4518
sdq13	0.8108	0.0693	0.1052	-0.1103	0.0086	0.2896	0.2791
sdq14	0.0006	0.4683	0.0882	-0.1813	0.4354	0.0278	0.4476
sdq15r	-0.0638	0.0974	0.0096	0.6782	-0.0056	-0.1264	0.4422
sdq16r	0.8193	-0.1041	-0.0755	0.0756	0.0652	-0.1517	0.2693
sdq17r	-0.0815	0.7545	-0.1584	0.0866	0.0177	-0.0457	0.4542
sdq18	-0.0213	-0.0627	0.9157	-0.0123	-0.0842	0.1863	0.1981
sdq19	0.8559	-0.0543	0.0220	-0.0275	-0.0445	-0.0764	0.3009
sdq20	0.0767	0.2417	0.1530	-0.1102	0.4084	-0.0761	0.6215
sdq21r	-0.0202	-0.0409	0.1534	0.7141	-0.0198	0.2510	0.4178
sdq22r	0.6240	-0.0685	-0.1181	0.3145	0.2459	0.3840	0.2746
sdq23r	0.0287	0.7989	-0.0257	-0.0692	-0.0369	0.0563	0.4214
sdq24	-0.0003	-0.0489	0.7555	0.0931	0.0431	-0.0753	0.3571
sdq25	0.7859	0.0011	0.0779	-0.1010	-0.0585	-0.0365	0.4217
sdq26	0.0447	0.1086	-0.0298	0.0353	0.5625	-0.0375	0.5986
sdq27r	0.0360	0.1530	0.3395	0.1177	-0.0193	-0.1785	0.7238
sdq28r	0.7811	0.0792	-0.1316	0.0970	0.0623	0.2432	0.2387
sdq29r	0.0400	0.7493	-0.1012	0.1272	0.1314	0.0413	0.2766
sdq30	-0.0373	-0.1128	0.7203	0.0937	0.2132	-0.0133	0.3322

Figure 16.13 Rotated solution for six factors

all three were originally intended to load on the verbal factor. Altogether, these four factors accounted for 56.4% of the total variance before rotation.

A standalone computer program entitled *Residuals* can be downloaded from edpsychassociates.com/Watkins3.html to ascertain model fit. The *Residuals* program requires the residual matrix as input (estat residuals command). That can be accomplished by copying the residual matrix from the Stata results with the ***Edit > Copy table*** menu options and then pasting it as text into an Excel file. That Excel file can then be saved as a text file for input to the *Residuals* program. The average overall residual misfit of this four-factor model was quantified by a root mean squared residual (RMSR) of .044. The smaller the RMSR value the better, with values ≤ .08 preferred (Brown, 2015) and values ≤ .05 indicating close fit (Maydeu-Olivares, 2017). Although the RMSR value was acceptable, a three-factor model might be more appropriate because there were only three salient loadings on the fourth factor with an alpha reliability (alpha sdq9r sdq15r sdq21r) of .62.

The three-factor model converged properly, produced reasonable parameter estimates, and accounted for 49.4% of the total variance before rotation. The average overall residual misfit of this model (RMSR) was .058, which is in the preferred range and approaches a close fit (Brown, 2015; Maydeu-Olivares, 2017).

```
Rotated factor loadings (pattern matrix) and unique variances
-------------------------------------------------------------------------
   Variable |  Factor1    Factor2    Factor3    Factor4 |  Uniqueness
------------+--------------------------------------------+---------------
       sdq1 |   0.8822    -0.0216    -0.0223    -0.1127  |   0.2463
       sdq2 |  -0.0792     0.7346    -0.0059    -0.0110  |   0.4999
       sdq3 |  -0.0203     0.1236     0.5239     0.3073  |   0.4848
      sdq4r |   0.6945    -0.0712    -0.1215     0.1770  |   0.4594
      sdq5r |   0.0312     0.6136    -0.1230     0.1901  |   0.5463
       sdq6 |   0.0854    -0.0213     0.6454    -0.1536  |   0.6096
       sdq7 |   0.7388     0.0065     0.1659    -0.2840  |   0.4419
       sdq8 |  -0.0534     0.6156     0.1928    -0.1000  |   0.5482
      sdq9r |   0.0334     0.0320     0.2561     0.5557  |   0.5470
     sdq10r |   0.5831     0.1215    -0.1501     0.2865  |   0.4388
     sdq11r |   0.0849     0.6134     0.0269     0.1374  |   0.4922
      sdq12 |  -0.0345    -0.0184     0.7213     0.0371  |   0.4739
      sdq13 |   0.8410     0.0522     0.1367    -0.2036  |   0.2812
      sdq14 |   0.0163     0.6977     0.1034    -0.1773  |   0.4973
     sdq15r |  -0.0555     0.1036     0.1288     0.6531  |   0.4723
     sdq16r |   0.7853    -0.0665    -0.1294     0.2076  |   0.3018
     sdq17r |  -0.0919     0.7572    -0.1923     0.1272  |   0.4714
      sdq18 |  -0.0275    -0.1371     0.9215    -0.0092  |   0.2215
      sdq19 |   0.8193    -0.0872    -0.0626     0.1043  |   0.3239
      sdq20 |   0.0754     0.4589     0.1510    -0.0469  |   0.7039
     sdq21r |   0.0505    -0.0285     0.4017     0.4042  |   0.6239
     sdq22r |   0.6964     0.0733     0.1196     0.0108  |   0.4625
     sdq23r |   0.0204     0.7515    -0.0836    -0.0337  |   0.4785
      sdq24 |  -0.0357    -0.0387     0.7214     0.2192  |   0.3939
      sdq25 |   0.7500    -0.0427    -0.0164     0.0220  |   0.4493
      sdq26 |   0.0800     0.4125     0.0723    -0.0026  |   0.7763
     sdq27r |  -0.0092     0.1352     0.2695     0.2769  |   0.7546
     sdq28r |   0.8210     0.1042    -0.0439    -0.0357  |   0.2662
     sdq29r |   0.0507     0.8160    -0.0788     0.1030  |   0.2825
      sdq30 |  -0.0468    -0.0021     0.7494     0.1443  |   0.3738
-------------------------------------------------------------------------
```

Figure 16.14 Rotated solution for four factors

As displayed in Figure 16.15, more than 80% of the salient loadings were good to excellent in magnitude (Comrey & Lee, 1992) in a simple structure configuration consistent with the theoretical underpinning of the scale, as color coded in Figures 16.11, although one item fell short of salience on the verbal factor (item 15r at .30). The interfactor correlations (estat common) of .01 to .44 were low enough to pose no threat to discriminant validity (Brown, 2015). Using the highest loadings on each factor, the alpha reliability coefficient was .91 for the math self-concept factor, .82 for the verbal self-concept factor, and .83 for the general self-concept factor.

This three-factor structure was robust to rotation method (promax and oblimin) as well as extraction method (iterated principal factor and maximum likelihood). Further, similar results were obtained when the potential outliers identified in Figure 16.5 were omitted. Finally, results did not differ when a Pearson correlation matrix was submitted to EFA and an oblique promax rotation applied.

```
Rotated factor loadings (pattern matrix) and unique variances

----------------------------------------------------------------
    Variable | Factor1    Factor2    Factor3 |  Uniqueness
-------------+------------------------------------+-------------
        sdq1 |  0.8754   -0.0596    -0.0585  |    0.2582
        sdq2 | -0.0907    0.7363    -0.0261  |    0.5117
        sdq3 | -0.0077    0.1740     0.6126  |    0.5007
       sdq4r |  0.7134   -0.0155    -0.0721  |    0.4912
       sdq5r |  0.0410    0.6841    -0.0856  |    0.5557
        sdq6 |  0.0518   -0.1055     0.5891  |    0.6987
        sdq7 |  0.6967   -0.0842     0.0737  |    0.5487
        sdq8 | -0.0715    0.5662     0.1543  |    0.6004
       sdq9r |  0.0707    0.1798     0.3963  |    0.7343
      sdq10r |  0.6068    0.2111    -0.0717  |    0.5071
      sdq11r |  0.0870    0.6570     0.0509  |    0.4896
       sdq12 | -0.0470   -0.0651     0.7396  |    0.4865
       sdq13 |  0.8119   -0.0187     0.0699  |    0.3483
       sdq14 | -0.0047    0.6241     0.0448  |    0.5857
      sdq15r | -0.0009    0.2773     0.2977  |    0.7614
      sdq16r |  0.8054   -0.0013    -0.0719  |    0.3457
      sdq17r | -0.0897    0.8194    -0.1782  |    0.4688
       sdq18 | -0.0508   -0.2037     0.9163  |    0.2739
       sdq19 |  0.8347   -0.0597    -0.0335  |    0.3304
       sdq20 |  0.0629    0.4321     0.1300  |    0.7242
      sdq21r |  0.0732    0.0651     0.5082  |    0.7003
      sdq22r |  0.6968    0.0620     0.1187  |    0.4622
      sdq23r |  0.0088    0.7522    -0.1120  |    0.4918
       sdq24 | -0.0316   -0.0341     0.7971  |    0.3854
       sdq25 |  0.7577   -0.0428    -0.0113  |    0.4455
       sdq26 |  0.0736    0.4057     0.0641  |    0.7825
      sdq27r |  0.0071    0.1957     0.3473  |    0.7799
      sdq28r |  0.8214    0.0912    -0.0607  |    0.2661
      sdq29r |  0.0487    0.8608    -0.0714  |    0.2774
       sdq30 | -0.0500   -0.0219     0.8034  |    0.3656
----------------------------------------------------------------
```

Figure 16.15 Rotated solution for three factors

Examination of these results suggested several plausible explanations. First, the verbal scale contained three items that addressed reading and seven items that dealt with English and written expression. Given the content of these items, children in elementary school may see reading and English as different subjects. Thus, a reading self-concept factor might emerge if more reading items were included. Additionally, the mixture of positively and negatively worded items may be responsible because they are known to disrupt the pattern of correlations and thereby create artifactual factors (Spector et al., 1997). That might be especially true for children in elementary school.

Additional support for the three-factor model was provided by analysis of a two-factor model. The two-factor model was marked by an RMSR of .09, with more than 20% of the residuals > .10. In this model, the math self-concept items continued to cohere but the verbal and general self-concept items collapsed into a single scale. This is a common symptom of underextraction and signals that the three-factor model appears to be the most acceptable solution.

A potential criticism of the scale in its entirety is the possibility of what Cattell (1978) called a "bloated specific." That is, items that are virtual paraphrases of each other. For example: "Do badly on math tests" and "Bad grades in math" might be perceived by young children as essentially the same question. Factors created by boated specifics are narrow and unlikely to replicate (de Winter et al., 2009). Likewise, factors created by combining items with similar vocabulary (e.g., math, reading, English) might be too narrow and unlikely to replicate (Gorsuch, 1997; Podsakoff et al., 2012). These sources of "hidden invalidity" should be considered (Hussey & Hughes, 2020, p. 166).

17 Higher-Order and Bifactor Models

Higher-Order Models

There are situations where constructs of different conceptual breadth exist and should, therefore, be reflected in the exploratory factor analysis (EFA) model. For example, the concept of general intelligence is currently thought to subsume narrower group factors which, in turn, subsume even narrower specific abilities (Carroll, 1993). This model reflects the theoretical belief that intelligence is comprised of subdimensions.

Broad, group, and specific abilities are often conceptualized in a higher-order factor model. In the simple higher-order model depicted in Figure 17.1, factors one through three are first-order factors (group factors) responsible for the intercorrelations of their indicator variables, whereas g is a broad general factor responsible for the intercorrelations of the first-order factors.

Gorsuch (1988) maintained that "correlated factors imply the existence of higher-order factors" (p. 250) and higher-order models have been advanced as a solution for the misinterpretation of rotated factors (Carretta & Ree, 2001). When factors are rotated, variance from the first factor is distributed across the remaining factors and seems to strengthen those factors. In actuality,

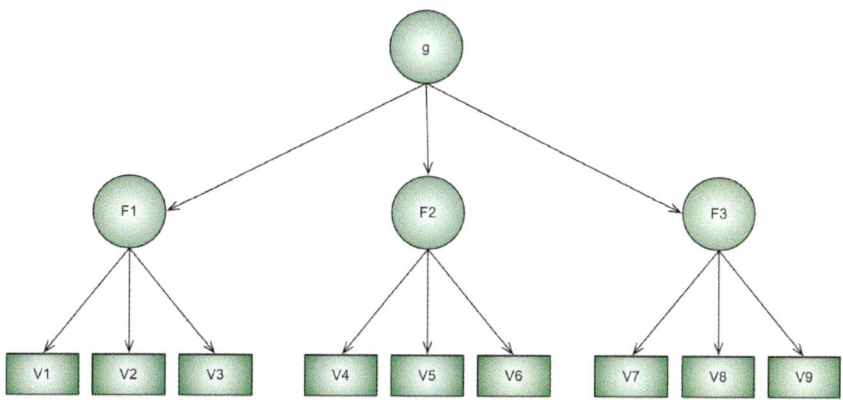

Figure 17.1 Simplified path diagram of a higher-order factor model

DOI: 10.4324/9781003149286-17

variance from the first factor has simply become a major source of variance in the new rotated factors. This can lead to a mistaken interpretation about the relative importance of the factors (Gignac, 2007). Similar opinions were expressed by Carroll (1983). Some researchers object to the use of EFA for higher-order analyses (Osborne & Banjanovic, 2016). However, EFA higher-order models are often encountered in the research literature (e.g., Canivez & Watkins, 2010; Watkins, 2006) and should be utilized if congruous with theory (Gorsuch, 1983; Thompson, 2004).

A second-order factor analysis can be conduced with the first-order factor correlation matrix as input and should be based on at least three first-order factors for identification. This requirement rules out the iq data used in the previous example. Likewise, the sdq data is not appropriate because there is no theoretical expectation of a broad general factor (Marsh, 1990). However, the classic Holzinger and Swineford (1939) data is often used to illustrate higher-order and bifactor models.

The HolzingerSwineford.xlsx file can be imported via the **File > Import > Excel spreadsheet** menu options. Once imported, the **Data Editor** window will display the data. Scroll through the data to ensure that there are nine variables and 301 participants. The `codebook` command will allow review of the characteristics of the data to ensure that all nine variables are numeric. There is no missing data indicator so it is assumed that there are no missing data. Once the data have been verified, they can be saved in Stata format for future convenience (**File > Save As > HolzingerSwineford**). Stata will recognize that this is a data file and automatically add the .dta suffix to the file name.

The `tabstat visper-speedcap, statistics (count mean sd min max skewness kurtosis)` command produces descriptive statistics that do not reveal any obviously incorrect or out-of-bounds values (Figure 17.2). The skew values were all near zero and the kurtosis values were all near three, indicating univariate normality. The `moments2` command can be used if zero-normed kurtosis values are desired. However, multivariate normality was marginal, given an expected kurtosis (Mardia, 1970) value of 99 and an obtained value of 103.6 ($X^2[1] = 8.0, p = .005$). There are 301 cases for all nine variables so there are no missing data.

As displayed in Figure 17.3 via the `correlate` command, the correlation matrix revealed that there are multiple coefficients $\geq .30$ (Hair et al., 2019; Tabachnick & Fidell, 2019).

stats	visper	cubes	lozenges	paracomp	sencomp	wordmean	speedadd	speeddot	speedcap
N	301	301	301	301	301	301	301	301	301
mean	4.93577	6.08804	2.250415	3.060908	4.340532	2.185572	4.185902	5.527076	5.374123
sd	1.167432	1.177451	1.130979	1.164116	1.290472	1.095603	1.089534	1.012615	1.009152
min	.6666667	2.25	.25	0	1	.1428571	1.304348	3.05	2.777778
max	8.5	9.25	9.25	4.5	6.333333	7	6.142857	7.434783	9.25
skewness	-.2556183	.4724289	.3853482	.2688252	-.3515465	.8622419	.2503346	.5278865	.2048911
kurtosis	3.329621	3.354647	2.106447	3.100695	2.463807	3.842043	2.710577	4.199412	3.311877

Figure 17.2 Descriptive statistics for Holzinger–Swineford data

	visper	cubes	lozenges	paracomp	sencomp	wordmean	speedadd	speeddot	speedcap
visper	1.0000								
cubes	0.2973	1.0000							
lozenges	0.4407	0.3398	1.0000						
paracomp	0.3727	0.1529	0.1586	1.0000					
sencomp	0.2934	0.1394	0.0772	0.7332	1.0000				
wordmean	0.3568	0.1925	0.1977	0.7045	0.7200	1.0000			
speedadd	0.0669	-0.0757	0.0719	0.1738	0.1020	0.1211	1.0000		
speeddot	0.2239	0.0923	0.1860	0.1069	0.1387	0.1496	0.4868	1.0000	
speedcap	0.3903	0.2060	0.3287	0.2078	0.2275	0.2142	0.3406	0.4490	1.0000

Figure 17.3 Correlation matrix for Holzinger–Swineford data

Based on results from the `factortest` `visper-speedcap` command (Figure 17.4), there is little likelihood of a multicollinearity problem with a determinant of .047. Bartlett's test of sphericity (1950) rejected the hypothesis that the correlation matrix was an identity matrix (chi-square of 904.1 with 36 degrees of freedom, $p < .001$). The Kaiser–Meyer–Olkin (KMO) measure of sampling adequacy was acceptable with a value of .75 (Hoelzle & Meyer, 2013; Kaiser, 1974). Altogether, these measures indicate that the correlation matrix is appropriate for EFA (Hair et al., 2019; Tabachnick & Fidell, 2019).

```
Determinant of the correlation matrix
Det                 =       0.047


Bartlett test of sphericity

Chi-square          =               904.097
Degrees of freedom  =                    36
p-value             =                 0.000
H0: variables are not intercorrelated


Kaiser-Meyer-Olkin Measure of Sampling Adequacy
KMO                 =       0.752|
```

Figure 17.4 Determinant, Bartlett test of sphericity, and KMO for Holzinger–Swineford data

Postestimation commands were activated following the `factor` `visper-speedcap`, `pcf` `factors(9)` command. First, the `screeplot` indicated that three factors would be sufficient (Figure 17.5), as did the parallel analysis `fapara`, `pca reps(500)` command where real eigenvalues are in blue and random eigenvalues in red. In contrast, the `minap` `visper-speedcap` command implementing the MAP criterion indicated that two factors would be sufficient.

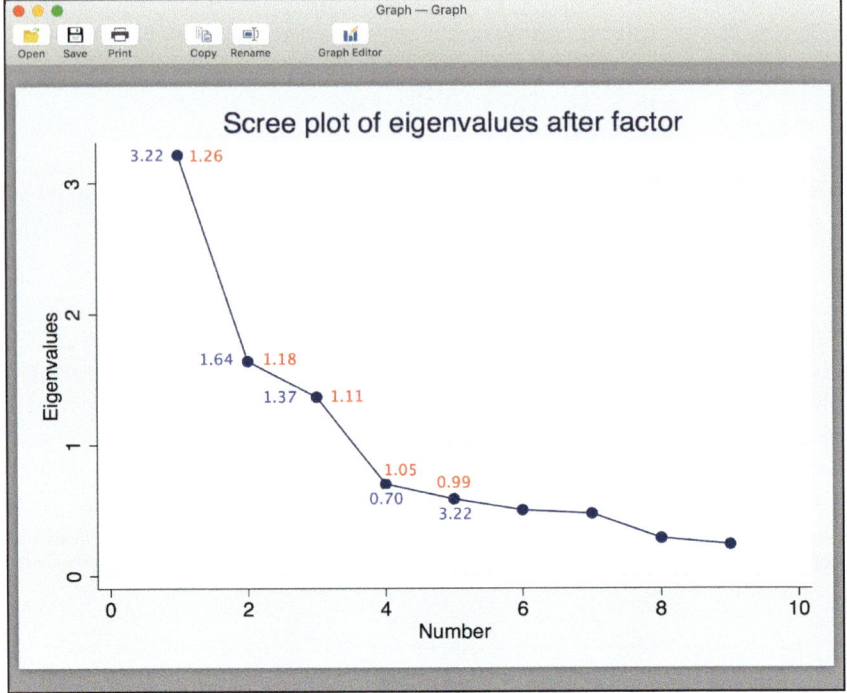

Figure 17.5 Scree plot with real (blue) and random (red) eigenvalues for Holzinger–
          Swineford data

```
Factor analysis/correlation                 Number of obs    =       301
Method: iterated principal factors          Retained factors =         3
Rotation: (unrotated)      |                Number of params =        24
                                                     Common Variance
-------------------------------------------------------------------------
     Factor  |  Eigenvalue  Difference  Total  Proportion   Cumulative
-------------+-----------------------------------------------------------
    Factor1  |    2.82748     1.61281   31.4%    0.5821       0.5821
    Factor2  |    1.21467     0.39926   13.5%    0.2501       0.8321
    Factor3  |    0.81541     0.70915    9.1%    0.1679       1.0000
    Factor4  |    0.10626     0.04882   -----    0.0219       1.0219
    Factor5  |    0.05744     0.05813   54.0%    0.0118       1.0337
    Factor6  |   -0.00069     0.01944            -0.0001      1.0336
    Factor7  |   -0.02013     0.02970            -0.0041      1.0294
    Factor8  |   -0.04983     0.04322            -0.0103      1.0192
    Factor9  |   -0.09305        .               -0.0192      1.0000
-------------------------------------------------------------------------
```

Figure 17.6 Unrotated solution for three factors

   An EFA with three factors and iterated principal factor extraction was
conducted with the `factor visper-speedcap, factors(3)`
`ipf` command. Those results are presented in Figure 17.6. Before rotation,
the first factor accounted for 31.4% of the total variance and 58% of the

common variance, the second factor for 13.5% of the total variance and 25% of the common variance, and the third factor for 9.1% of the total variance and 17% of the common variance. Combined, the three factors accounted for 54% of the total variance.

Promax rotation was accomplished with the `rotate,    promax` command. As anticipated, three factors emerged: verbal ability, spatial ability, and mental speed (Figure 17.7). Although one of the mental speed items also saliently loaded on the spatial factor, the three-factor solution appears to be adequate. Structure coefficients were generated with the `estat structure` command and were strong (.49 to .87) without evidence of a suppression effect (Thompson, 2004).

```
Rotated factor loadings (pattern matrix) and unique variances
-------------------------------------------------------------
    Variable | Factor1    Factor2    Factor3 |  Uniqueness
-------------+-------------------------------+---------------
      visper |  0.1651     0.6057     0.0173 |    0.5233
       cubes |  0.0171     0.5218    -0.1344 |    0.7448
    lozenges | -0.0989     0.7033     0.0032 |    0.5465
    paracomp |  0.8462     0.0140     0.0071 |    0.2721
     sencomp |  0.8900    -0.0690     0.0080 |    0.2463
    wordmean |  0.8021     0.0791    -0.0150 |    0.3086
    speedadd |  0.0458    -0.1579     0.7416 |    0.4818
    speeddot | -0.0464     0.1266     0.6856 |    0.4796
    speedcap |  0.0078     0.3906     0.4487 |    0.5395
-------------------------------------------------------------
```

Figure 17.7 Rotated solution for three factors

Analysis of the residual matrix (`estat residuals`) revealed a close fit with an RMSR of .019 with no residuals greater than .05 (Maydeu-Olivares, 2017). In sum, the three-factor model appears to be a good EFA solution that offers the optimum balance between comprehensiveness (accounting for the most variance) and parsimony (with the fewest factors).

Although the three-factor model appeared to be satisfactory, a two-factor model was also generated with the `factor    vocab1-designs2,    factors(2) ipf` command to ensure that it did not exhibit superior fit characteristics. This model accounted for 43.4% of the total variance before rotation. Following oblique rotation, the variables that previously loaded on the spatial and mental speed factors collapsed onto a single factor, signaling underextraction. Additionally, the RMSR value was .084 and almost 20% of the residuals exceeded .10. Thus, measures of model fit remove this model from consideration, leaving the three-factor model the preferred solution.

However, theory (Carroll, 1993; Gorsuch, 1988) suggests that there is an overarching general factor that is responsible for the intercorrelations between these three first-order factors as demonstrated by moderate interfactor correlations generated by the `estat common` command (Figure 17.8).

```
Correlation matrix of the promax(3) rotated common factors

----------------------------------------------------------------
    Factors |   Factor1    Factor2    Factor3
------------+---------------------------------------------------
    Factor1 |      1
    Factor2 |    .3745         1
    Factor3 |    .2229      .2929         1
----------------------------------------------------------------
```

Figure 17.8 Correlation matrix of the three rotated common factors

A higher-order EFA can be conducted by copying the interfactor correlation matrix into a syntax command file and then conducting an EFA on that factor intercorrelation matrix as was illustrated in Figure 4.5. That modified do-file is displayed in Figure 17.9. Note that with only one factor there is no need for rotation.

```
InputCorrMatHO.do
Open  Save  Print      Find  Show  Zoom                                    Do
1   * Input correlation matrix
2   * Use /// to indicate that line continues
3   matrix C = ( 1.00,0.3745,0.2229 \ ///
4   0.3745,1.00,0.2929 \ ///
5   0.2229,0.2929,1.00 )
6   * EFA commands follow
7   factormat C, n(301) names (verbal spatial speed) ///
8   fac(1) ipf
9
Automatic    Line: 9, Col: 1
```

Figure 17.9 Do-file for higher-order EFA from the factor intercorrelation matrix

Figure 17.10 reveals that the higher-order factor accounted for 31.7% of the total variance, with each first-order factor saliently loading on the general factor (.42 to .70).

These two analyses can be summarized in a path diagram that displays both first- and second-order loadings (Figure 17.11). This model is conceptually more complicated than a first-order model because each measured variable is influenced by two factors, one of them far removed from the measured variables and thus more abstract and difficult to understand (Hair et al., 2019). Thompson (2004) suggested the analogy of a mountain range: first-order factors provide a close-up view that focuses on the details of the valleys and the peaks whereas the second-order factors are like looking at the mountains at a great distance, seeing them as constituents of a range. Alternatively, first- and second-order factors might be compared to viewing a mountain range through a camera's telephoto and wide-angle lenses, respectively (McClain, 1996). These "perspectives complement each

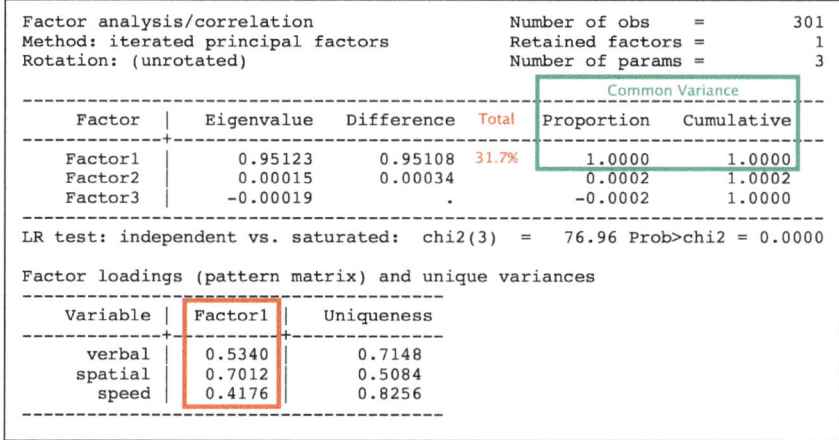

```
Factor analysis/correlation                    Number of obs   =        301
Method: iterated principal factors             Retained factors =         1
Rotation: (unrotated)                          Number of params =         3
                                                         Common Variance
---------------------------------------------------------------------------
    Factor  |  Eigenvalue   Difference  Total  Proportion   Cumulative
------------+--------------------------------------------------------------
    Factor1 |     0.95123     0.95108   31.7%    1.0000       1.0000
    Factor2 |     0.00015     0.00034            0.0002       1.0002
    Factor3 |    -0.00019        .              -0.0002       1.0000
---------------------------------------------------------------------------
LR test: independent vs. saturated:  chi2(3)  =   76.96 Prob>chi2 = 0.0000

Factor loadings (pattern matrix) and unique variances
-------------------------------------------------------
   Variable |  Factor1  |  Uniqueness
------------+-----------+-------------
     verbal |   0.5340  |   0.7148
    spatial |   0.7012  |   0.5084
      speed |   0.4176  |   0.8256
-------------------------------------------------------
```

Figure 17.10 Higher-order factor solution

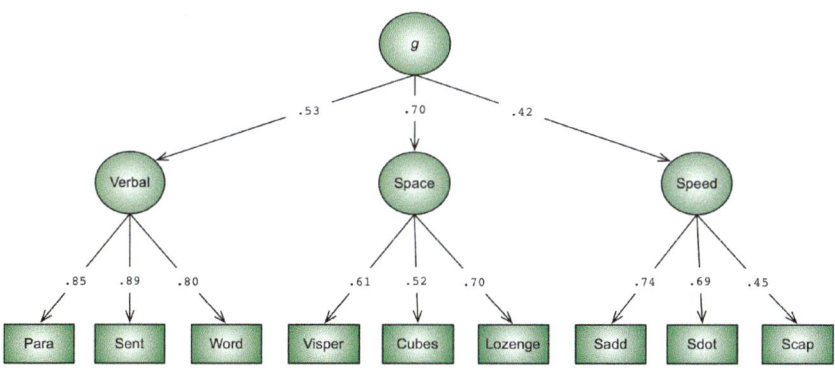

Figure 17.11 Path diagram of higher-order factor model for Holzinger–Swineford data

other, and each is needed to see patterns at a given level of specificity versus generality" (Thompson, 2004, p. 73).

The primary problem with higher-order models is interpretation because the first-order factor loadings represent two sources of variance (from both the general factor and the group factor). Thus, the relationship between a second-order factor and a measured variable is fully mediated by the first-order factor. The researcher must, therefore, provide a theoretical justification for the full mediation implied by the higher-order model (Gignac, 2008) and must consider the multiple sources of variance when interpreting the factors and measured variables.

Schmid–Leiman Transformation of Higher-Order Models

"Factors are abstractions of measured variables. Second–order factors, then, are abstractions of abstractions even more removed from the measured variables. Somehow we would like to interpret the second–order factors in terms of the measured variables, rather than as a manifestation of the factors of the measured variables" (Thompson, 2004, p. 74). It is possible to disentangle the variance due to general and group factors and identify the effect of a second–order factor on a measured variable by multiplying the path coefficients (e.g., .53 x .85 = .45; .53 x .89 = .47, etc.). A more elegant solution was provided by Schmid and Leiman (1957) who demonstrated that a higher–order model could be transformed into "orthogonal sources of variance: (1) variance shared between the higher–order general factor and the observed variables; and (2) variance shared between the first–order factors and the respective observed variables specified to load upon them" (Gignac, 2007, p. 40). Orthogonal (uncorrelated) factors are conceptually simpler to interpret because they are independent sources of variance (Carroll, 1983).

Two advantages of the Schmid–Leiman transformation (S–L) are "the calculation of direct relations between higher–order factors and primary variables, and the provision of information about the independent contribution of factors of different levels to variables" (Wolff & Preising, 2005, p. 48). The S–L was extensively employed by Carroll (1993) and has been recommended by other methodologists (Cattell, 1978; Gignac, 2007; Gorsuch, 1983; Humphreys, 1982; Lubinski & Dawis, 1992; McClain, 1996; Thompson, 2004).

The S–L transformation is not included in Stata. However, a standalone computer program named MacOrtho that computes the S–L transformation

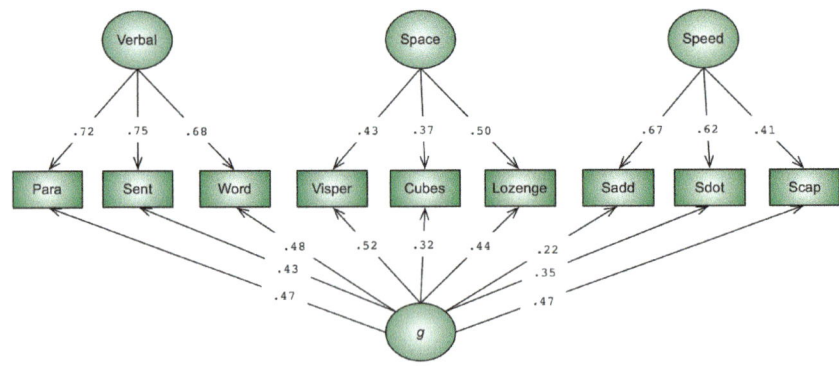

Figure 17.12 Path diagram of Schmid–Leiman results from MacOrtho program for Holzinger–Swineford data

can be downloaded from http://edpsychassociates.com/Watkins3.html. That program requires that the first-order pattern matrix, second-order loadings, variable names, and factor names be entered. Placing these factor loadings into a path diagram as in Figure 17.12 visually portrays the orthogonal relationships of group and general factors to the measured variables.

Bifactor Models

Holzinger and Swineford (1937) proposed the bifactor model where each observed variable depends on two factors: a general factor and a smaller group factor characterizing a specific subset of the measured variables. The general factor has direct effects on all measured variables but not on the group factors. Thus, it has wide breadth (Humphreys, 1982). The group factors have direct effects on subsets of measured variables and therefore narrower breadth. Both the general and group factors are uncorrelated, allowing for clarity of interpretation. "Bifactor models are potentially applicable when (a) there is a general factor that is hypothesized to account for the commonality of the items; (b) there are multiple domain specific factors, each of which is hypothesized to account for the unique influence of the specific domain over and above the general factor; and (c) researchers may be interested in the domain specific factors as well as the common factor that is of focal interest" (Chen et al., 2006, p. 190). Reise (2012) argued that a bifactor model, "which views the variance in trait indicators as being influenced by both general and group sources of variance, provides a strong foundation for understanding psychological constructs and their measurement" (p. 692).

Bifactor models appear identical to Schmid–Leiman transformed models when illustrated in a path diagram. However, bifactor models are not mathematically equivalent to the S–L transformation of higher-order models (Chen et al., 2006). Rather, the transformed higher-order model includes mediating variables and proportionality constraints that may bias the estimation of population values (Mansolf & Reise, 2016; Reise et al., 2010). Accordingly, Mansolf and Reise (2016) said to "treat this method [S–L] as a descriptive technique only, and not as an estimator of a bifactor structure in the population" (p. 714).

Nevertheless, exploratory bifactor models have typically been estimated by S-L transformations because software to compute an exploratory bifactor analysis was not readily available (Reise, 2012). That deficiency was remedied by Jennrich and Bentler (2011) who explicated the mathematics of an exploratory bifactor rotation that was subsequently included in the *psych* package within the **R** (R Core Team, 2020) statistical system. It is not included in Stata.

There has been little research on the properties of Jennrich and Bentler's (2011) exploratory bifactor rotation method but Mansolf and Reise (2015, 2016) reported that it produced biased parameter estimates in some conditions and was vulnerable to local minima (i.e., erroneous parameter estimates). More recently, it was found to be sensitive to variables with small factor loadings and relative large cross-loadings (Dombrowski et al., 2019) and less accurate than Schmid–Leiman transformations in recovering population structures (Giordano & Waller, 2020). Given these results, exploratory bifactor analysis should be employed with care. More research is needed to clarify the strengths and weaknesses of exploratory bifactor analysis rotations (Lorenzo-Seva & Ferrando, 2019). Before employing higher-order or bifactor models, the user should consult Brunner et al. (2012), Chen et al. (2006), Mansolf and Reise (2016), and Reise (2012).

Alternative Measures of Reliability

To this point, reliability estimation has consisted of coefficient alpha (Cronbach, 1951). Although popular, most applications of coefficient alpha have ignored its statistical assumptions, resulting in biased estimates of reliability (Watkins, 2017). Model-based reliability estimates have been proposed as alternatives to alpha that make fewer and more realistic assumptions (Reise, 2012). Especially useful are the omega (ω) family of coefficients described by McDonald (1999). Based on an orthogonal factor model, these indices allow a judgment of the relative strength of the general factor and its influence on derived scale scores as well as the viability of global and subscale domains (Rodriguez et al., 2016; Zinbarg et al., 2005). Tutorials on model-based estimates of reliability were provided by Watkins (2017) and Flora (2020).

Several omega variants can be computed to describe how precisely "total and subscale scores reflect their intended constructs" and determine "whether subscale scores provide unique information above and beyond the total score" (Rodriguez et al., 2016, p. 223). The most general omega coefficient is omega total (ω), "an estimate of the proportion of variance in the unit-weighted total score attributable to all sources of common variance" (Rodriguez et al., 2016, p. 224). Omega total is essentially a model-based substitute for alpha. Stata does not contain routines for computing omega reliability estimates but an ado-file to estimate omega total can be installed with the `ssc install omega` command and implemented with the `omega visper-speedcap` command. This resulted in an omega estimate of .72 for a single scale score. In contrast, the `alpha visper-speedcap` command resulted in an alpha estimate of .76. In agreement with Hayes and Coutts (2020), alpha and omega "typically produce similar estimates of reliability when applied to real data" (p. 20).

An omega total coefficient can also be estimated for each group factor. That is, a model-based substitute for coefficient alpha in that they represent the ratio of variance accounted for by all common factors (general plus group) to total variance. These can be implemented with the `omega paracomp sencomp wordmean`, `omega visper cubes lozenges`, and `omega speedadd speeddot speedcap` commands and produce coefficients of .89, .64, and .70 for verbal, spatial, and speed scales, respectively.

Using the guidelines for alpha coefficients, only the reliability of the verbal scale is acceptable given the group nature of this research but it might not be sufficient for making important decisions about individuals (DeVellis, 2017).

However, the amalgam of general and group factor variance reflected by ω total does not allow the contributions of general and group factor variance to be disentangled. Another omega variant, called omega hierarchical (ω^h), reflects the proportion of systematic variance in a unit-weighted score that can be attributed to the focal factor alone. A final omega variant is called omega hierarchical subscale (ω^{hs}), which indicates the proportion of variance in the subscale score that is accounted for by its intended group factor to the total variance of that subscale score and indexes the reliable variance associated with that subscale after controlling for the effects of the general factor.

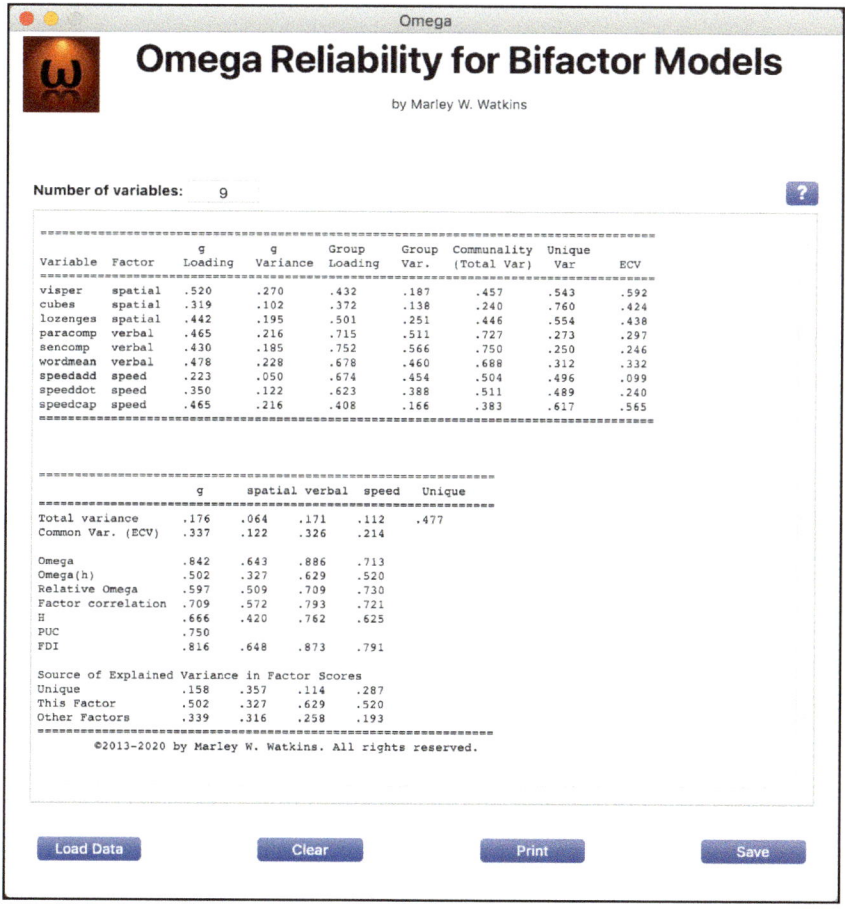

Omega Reliability for Bifactor Models

by Marley W. Watkins

Number of variables: 9

Variable	Factor	g Loading	g Variance	Group Loading	Group Var.	Communality (Total Var)	Unique Var	ECV
visper	spatial	.520	.270	.432	.187	.457	.543	.592
cubes	spatial	.319	.102	.372	.138	.240	.760	.424
lozenges	spatial	.442	.195	.501	.251	.446	.554	.438
paracomp	verbal	.465	.216	.715	.511	.727	.273	.297
sencomp	verbal	.430	.185	.752	.566	.750	.250	.246
wordmean	verbal	.478	.228	.678	.460	.688	.312	.332
speedadd	speed	.223	.050	.674	.454	.504	.496	.099
speeddot	speed	.350	.122	.623	.388	.511	.489	.240
speedcap	speed	.465	.216	.408	.166	.383	.617	.565

	g	spatial	verbal	speed	Unique
Total variance	.176	.064	.171	.112	.477
Common Var. (ECV)	.337	.122	.326	.214	
Omega	.842	.643	.886	.713	
Omega(h)	.502	.327	.629	.520	
Relative Omega	.597	.509	.709	.730	
Factor correlation	.709	.572	.793	.721	
H	.666	.420	.762	.625	
PUC	.750				
FDI	.816	.648	.873	.791	

Source of Explained Variance in Factor Scores				
Unique	.158	.357	.114	.287
This Factor	.502	.327	.629	.520
Other Factors	.339	.316	.258	.193

| Load Data | Clear | Print | Save |

Figure 17.13 Omega reliability output from Omega program for Holzinger–Swineford data

There are no Stata EFA routines for computation of ω^h or ω^{hs} . However, a standalone program called Omega can be downloaded from http://edpsychassociates.com/Watkins3.html to compute those metrics. The first- and second-order loadings in Figure 17.12 serve as input to the Omega program. There will be some differences in the values produced by the Omega command and the Omega standalone program due to computation method, use of a Schmid–Leiman model, and roundoff error (Garcia-Garzon et al., 2020; Giordano & Waller, 2020).

The results of those computations are presented in Figure 17.13. Omega total estimates for the general, verbal, spatial, and speed factors were .84, .89, .64, and .71, respectively. Omega hierarchical subscale (ω^{hs}) coefficients reflects the reliable variance associated with that subscale after controlling for the effects of the general factor. In this case, ω^{hs} values of .63, .33, and .52 were estimated for the verbal, spatial, and speed scales, respectively.

There is no universally accepted guideline for acceptable or adequate levels of omega reliability for clinical decisions, but it has been recommended that omega hierarchical coefficients should exceed .50 at a minimum with .75 preferable (Reise, 2012). Additional details about the use and interpretation of omega coefficients are provided by Rodriguez et al. (2016), Watkins (2017), and Zinbarg et al. (2005). A tutorial on use of the Omega program is also available (Watkins & Canivez, 2021).

18 Exploratory versus Confirmatory Factor Analysis

Exploratory (EFA) and confirmatory factor analysis (CFA) are both based on the common factor model and both attempt to reproduce the correlations among a set of measured variables with a smaller set of latent variables (Brown, 2015). However, the two methods differ in the measurement model: EFA allows all factors to relate to all measured variables whereas CFA restricts relationships between factors and measured variables to those specified beforehand by the analyst. Consequently, EFA is sometimes called unrestricted factor analysis and CFA is called restricted factor analysis (Widaman, 2012).

EFA and CFA models are represented by the path diagrams in Figure 18.1. In the EFA model, every variable loads onto every factor. The researcher must apply the principle of simple structure and specify the magnitude of factor loadings required for salience to remove the presumably trivial variable–factor relationships from consideration. In CFA, the researcher must specify beforehand the number of factors, how the factors relate to each other, and which variables load onto which factor. In a typical CFA (as depicted in the CFA model), the loadings of variables V1–V3 onto Factor 2 and those of variable V4–V6 onto Factor 1 are prespecified to be zero and therefore omitted from the path diagram. Although other patterns are possible, this independent clusters model (ICM) is most common (each variable is allowed to load on one factor and its loadings on all other factors

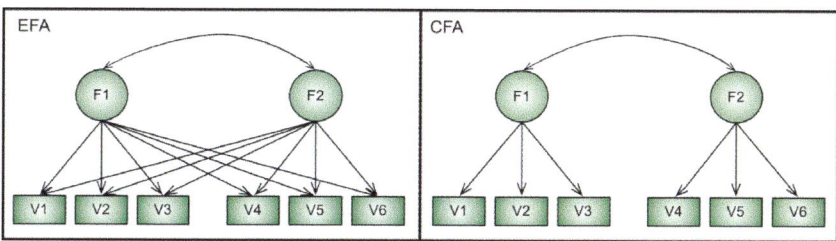

Figure 18.1 Simplified path diagrams of EFA and CFA models

DOI: 10.4324/9781003149286-18

constrained to zero). This a priori specification of simple structure removes the need for factor rotation in CFA.

Many methodologists believe that "the purpose of EFA is typically to identify the latent constructs or to generate hypotheses about their possible structures, whereas the purpose of CFA is to evaluate hypothesized structures of the latent constructs and/or to develop a better understanding of such structures" (Bandalos & Finney, 2019, p. 98). That is, EFA generates models whereas CFA tests models (Matsunaga, 2010; Norman & Streiner, 2014). Obviously, any CFA model's veracity depends on the accuracy and thoroughness of its prespecified parameters. As judged by Carroll (1993), "it is highly desirable, or perhaps mandatory, that the hypotheses to be tested have excellent logical or psychological support in some theory of individual differences and their measurements, or in prior analyses of datasets other than the one on which the hypotheses are to be tested" (p. 82).

Some methodologists consider EFA to be inferior to CFA because EFA "tends to be guided by ad hoc rules and intuition, whereas confirmatory factor analysis is imbued with robust dual tests for both parameters (factor loadings) and the quality of the factor" (Ramlall, 2017, p. 5). Likewise, Little et al. (2017) disparaged EFA, asserting that "EFA is a data-driven enterprise that presents too many misguiding temptations for misunderstanding the nature of the construct of interest. Using CFA with careful and skillful scrutiny of all indicated modifications will reveal the structure of a scale that will retain optimal validity" (p. 127).

CFA Model Fit

Thus, it is often claimed that CFA provides an objective test of factor models (Hoyle, 2000; Ramlall, 2017). Maximum likelihood estimation is typically employed in CFA (Russell, 2002) and CFA models are commonly assessed with a statistical test of overall model fit, the chi-square goodness-of-fit, that assesses the discrepancy between the observed relationships among the measured variables versus the relationships implied by the prespecified model. The chi-square value will be statistically non-significant if the relationships are identical but statistically significant if the model and actual data relationships differ (Barrett, 2007). Unfortunately, the chi-square test is sensitive to sample size: large samples may produce statistically significant chi-square values even when the difference between actual and model implied relationships is trivial and small sample sizes may fail to distinguish between quite discrepant models (Kline, 1994).

In reality, most CFA models produce statistically significant chi-square values so common practice is to ignore that result and rely on approximate fit indices that measure some aspect of closeness of fit of the actual and implied models (Brown, 2015; Browne, 2001; Hancock & Schoonen, 2015; Hoyle, 2000; Hurley et al., 1997; Maydeu-Olivares, 2017; Ropovik, 2015; Saris et al., 2009). Essentially, this practice acknowledges that the model is not an exact representation of reality but that it might be useful if it sufficiently corresponds to the real world and is not entirely incorrect (MacCallum, 2003). Thus, the idea that CFA provides an objective test of models is both true and false: true that the chi-square test of exact fit is conducted but false that the model is disconfirmed by that test in typical practice (Hurley et al., 1997; Ropovik, 2015). Rather, approximate fit indices are used to subjectively judge that a model is close enough to reality to be useful.

Unfortunately, there is no clear empirical demarcation to the quantification of "close enough" (Berk, 2011). Dozens of approximate fit indices have been developed to measure a model's "closeness to reality" but many "are subjective in nature or do not follow known statistical distributions" (DiStefano, 2016, p. 167). In practice, fit indices may conflict (i.e., one signaling close fit and another poor fit); thus, "researchers are likely to be confused and potentially make incorrect research conclusions" (Lai & Green, 2016, p. 220). Statistical simulations have been conducted to develop "rules of thumb" for optimal cut-points of approximate fit indices (e.g., Hu & Bentler, 1999) but each fit index is differentially sensitive to estimation method, sample size, distributional characteristics, model size, and model misspecification (Clark & Bowles, 2018; Preacher & Merkle, 2012; Saris et al., 2009; Savalei, 2012; Shi & Maydeu-Olivares, 2020; Tomarken & Waller, 2003) with the result that "cutoff values for fit indices, confidence intervals for model fit/misfit, and power analysis based on fit indices are open to question" (Yuan, 2005, p. 142). As summarized by Greiff and Heene (2017), "there are no golden rules for cutoff values, there are only misleading ones" (p. 315).

Approximate fit indices are not sensitive in detecting the correct number of factors (Heene et al., 2011), especially with ordinal data (Garrido et al., 2016; Themessl-Huber, 2014). Nor do they uniquely identify correctly specified models. This was vividly illustrated by Orcan (2018), who simulated a simple model with two correlated factors, each with four measured variables. CFA fit indices indicated good fit for the correct model as well as for two purposively misspecified models. In common practice, "if a goodness-of-fit index meets the recommended cutoff values, the model is retained and used as if it were correctly specified ... [and] inferences on the model parameters are made as if the model were correct, when in fact, it is misspecified" (Maydeu-Olivares, 2017, pp. 538–539). By relying on goodness-of-fit, "essentially unidimensional measures might require multidimensional specifications for the degree of fit to be acceptable. And the resulting solution generally consists of minor, ill-defined factors that yield poor factor score estimates" (Ferrando et al., 2019, p. 124).

Browne and Cudeck (1993) reminded researchers that "fit indices should not be regarded as measures of usefulness of a model. They contain some information about the lack of fit of a model, but none about plausibility" (p. 157). Even worse, subjectivity in model selection is enhanced by the availability of numerous fit indices, each with different cut-points (Hayduk, 2014; Heene et al., 2011; McCoach et al., 2013; McDonald, 2010). The probability that "close enough" models are scientifically useful is usually neglected, but may be unlikely according to basic principles of logic and science (Meehl, 2006; Platt, 1964; Popper, 2002; Roberts & Pashler, 2000; Ropovik, 2015; Wasserstein & Lazar, 2016).

CFA Model Respecification

If a model with acceptable fit cannot be found among the candidate models, model modifications or respecifications are often guided by "modification indices" produced by the CFA software (MacCallum, 2003; Marsh et al., 2009; Maydeu-Olivares, 2017; McCoach et al., 2013; Tarka, 2018. Modification indices statistically approximate how much the chi-square goodness of fit test and accompanying approximate fit indices might improve if a particular prespecified fixed parameter (i.e., factor loading, interfactor correlation, etc.) were allowed to be estimated rather than fixed (Brown, 2015). According to Brown (2015), model modifications "should only be pursued if they can be justified on empirical or conceptual grounds" (Brown, 2013, p. 265). However, "if a model modification is highly interpretable, then why was it not represented in the initial model" (MacCallum et al., 1992, p. 492)? In fact, it does not seem difficult for researchers to find theoretical justifications for almost any respecification (Armstrong & Soelberg, 1968; Steiger, 1990). Additionally, respecifications tend not to generalize to new samples nor to the population (Bandalos & Finney, 2019; Kline, 2012, 2013; MacCallum et al., 1992), especially when there are fewer than 1,200 participants in the sample (Hutchinson, 1998). The respecification procedure has been called "a dangerous practice because it capitalizes on chance … none of the significance tests nor the goodness-of-fit measures take this capitalization into account" (Gorsuch, 2003, p. 151). Almost identical warnings have been sounded by other methodologists (Cliff, 1983; Cribbie, 2000; Jebb et al., 2017; McArdle, 2011; Sass & Schmitt, 2010; Tomarken & Waller, 2003). If not disclosed, respecification based on examination of the data is unethical and can impede scientific progress (Breckler, 1990; Murphy & Aguinis, 2019; Rubin, 2017; Simmons et al., 2011).

Box (1976) provided cogent scientific guidance more than 40 years ago. "Since all models are wrong the scientist cannot obtain a 'correct' one by excessive elaboration. On the contrary following William of Occam he should seek an economical description of natural phenomena … overelaboration and overparameterization is often the mark of mediocrity" (p. 792). Therefore, even if justified by theory, respecification means that the analysis has become exploratory rather than confirmatory (Brown, 2015; Browne, 2001; Gorsuch, 1997; Hancock & Schoonen, 2015; Ropovik, 2015; Schmitt, 2011; Tarka, 2018) and "has the worst of both exploratory and confirmatory factor analysis and cannot be recommended" (Gorsuch, 1988, p. 235). Accordingly, "it is better to identify potential problems [via EFA] prior to fitting a confirmatory model than it is to fit a confirmatory model, find problems in fit, and try to diagnose (and 'fix') them" (Reise et al., 2018, p. 699).

CFA Model Assumptions

These complexities are often overlooked by researchers (Hayduk, 2014; Hoyle, 2000; Ramlall, 2017; Schmitt et al., 2018). Additionally, researchers tend to ignore the statistical assumptions necessary for unbiased CFA results (Kline, 2012) and "the sheer technical complexity of the [CFA] method tends to overwhelm critical judgement" (Freedman, 1987, p. 102). Kline (2012) reported that several assumptions must be considered in CFA. These include attentiveness to the direction of causal influence from factor to measured variables and the absence of other plausible explanations.

At the most basic level, it is assumed that the CFA model is correctly specified (DiStefano, 2016; Kline, 2012). Three primary model specifications are assumed: (a) the correct number of factors, (b) the correct pattern of factor loadings, and (c) the absence of unmodeled subfactors (Hoyle, 2000). Assumptions about the data in CFA include those assumptions that underlie use of maximum likelihood estimation applied to the covariance, rather than the correlation, matrix (Brown, 2013; Kline, 2012). Specifically, "linearity, independent observations, sufficiently large sample size, multivariate normality of indicators, and a correctly specified model" (DiStefano, 2016, p. 167). Thus, CFA results will be unbiased if the measured variables are multivariate normally distributed, the number of participants is sufficiently large, and the model is correctly specified (DiStefano, 2016). However, correct model specification is unlikely in many practical situations (Bandalos, 2018; DiStefano, 2016; Hurley et al., 1997; Kline, 2012; Orcan, 2018; Tomarken & Waller, 2003, 2005). In fact, CFA has been found to be especially prone to overfactoring with ordinal data (Garrido et al., 2016; van der Eijk & Rose, 2015; Xia & Yang, 2019). Additionally, data in the social sciences are rarely multivariate normally distributed (Cain et al., 2017) and "all inferences based on [misspecified models] are suspect" (Maydeu-Olivares, 2017, p. 533), consequently "the interpretation of results in the typical [CFA] may be unwarranted" (Kline, 2012, p. 111).

The limitations of CFA have become especially apparent to researchers who investigate the structure of psychological instruments (Hoelzle & Meyer, 2013; Hopwood & Donnellan, 2010; Marsh et al., 2009; Wetzel & Roberts, 2020). The typical independent clusters model (ICM) approach appears to be too restrictive because test items are often ordinal and exhibit small cross-loadings on several factors (Hurley et al., 1997; Schmitt, 2011). Thus, some items may manifest localized weak discriminant validity (i.e., equivalent loadings on multiple factors) that is hidden within the overall CFA model fit. In an EFA, these small cross-loadings are assumed to be indicators of ill-defined factors of no substantive interest and are therefore ignored. However, constraining these minor loadings to zero in CFA causes a degradation of model fit and systematic inflation of the interfactor correlations. For example, Marsh et al. (2009) found that a median correlation of .34 among factors in an EFA increased to .72 in a CFA. These artificially inflated factor correlations may negatively impact subsequent efforts to identify unique external correlates of the factors (Molina et al., 2020). As summarized by Schmitt et al. (2018), violation of the ICM assumption can lead to "rather arbitrary modifications, unrealistic factor loadings, and elevated interfactor correlations" (p. 349). However, an EFA model will be unaffected by these conditions. See Morin et al. (2020) for an in-depth discussion of this issue.

Exploratory Use of CFA

Reliance on approximate fit indices, post hoc modification of models based on data characteristics, and violation of statistical assumptions have created a situation where CFA is often used in an exploratory manner (Bandalos & Finney, 2019; Browne, 2001; Cliff, 1983; DiStefano, 2016; Gerbing & Hamilton, 1996; Gorsuch, 1988; Hancock & Schoonen, 2015; Heene et al., 2011; Kline, 2012; MacCallum et al., 1992; Maydeu-Olivares, 2017; Tomarken & Waller, 2003, 2005). Some methodologists have suggested that CFA is superior to EFA because it provides a statistical test of models and its decisions are more objective (Little et al., 2017; Ramlall, 2017). However, exploratory use of CFA leaves researchers vulnerable to confirmatory bias (MacCallum & Austin, 2000; Roberts & Pashler, 2000). That is, they may "persevere by revising procedures until obtaining a theory-predicted result" (Greenwald et al., 1986).

As previously described, statistical tests in CFA are conducted but often ignored (Hayduk, 2014). Likewise, the decision steps in CFA are not nearly as objective as claimed: "with all its breadth and versatility come many opportunities for misuse" (Hancock & Schoonen, 2015, p. 175). For example, DiStefano and Hess (2005) reviewed published CFA studies and found that more than 50% provided inadequate theoretical support for the tested models, failed to consider statistical assumptions of CFA, and appeared to select approximate fix indices and their thresholds opportunistically. Literature reviews have also found that published CFA studies often fail to report critical information such as method of estimation, type of matrix analyzed, justification of sample size, and distributional characteristics of the data (MacCallum & Austin, 2000; Ropovik, 2015; Russell, 2002).

EFA versus CFA

Given the increased popularity of CFA methods and the widespread belief that EFA is not acceptable, researchers may be criticized by journal editors for choosing EFA over CFA (Haig, 2018; Hayduk, 2014; Hurley et al., 1997; Marsh et al., 2009). However, the differences between EFA and CFA are not as distinct as sometimes claimed (Child, 2006; Fabrigar & Wegener, 2012; Hoelzle & Meyer, 2013; Hoyle, 2000). EFA and CFA are both valuable statistical tools for scientific inquiry (Gorsuch, 1988). Haig (2018) discussed the role of factor analysis in scientific method and argued that "science is as much concerned with theory generation as it is with theory testing" (p. 84) and that "EFA has a legitimate, and important, role as a method of

theory generation, and that EFA and CFA should be viewed as comple-
mentary, not competing, methods of common factor analysis" (p. 83). Clark
and Bowles (2018) noted that to the extent that EFA and CFA "methods
converge on a conceptually coherent solution, it is possible to have greater
confidence in that solution, even if there are documented flaws with any one
method" (p. 555). Other methodologists have also suggested that EFA can
be productively used in conjunction with CFA (Bandalos & Finney, 2010;
Carroll, 1995a; DeVellis, 2017; Fabrigar et al., 1999; Gerbing & Hamilton,
1996; Goldberg & Velicer, 2006; Gorsuch, 2003; Hurley et al., 1997; Jebb
et al., 2017; Morin et al., 2016; Nesselroade, 1994; Schmitt, 2011; Schmitt
et al., 2018; Sellbom & Tellegen, 2019; Wegener & Fabrigar, 2000).

In summary, both EFA and CFA require thoughtful and evidence-based
methodological decisions that offer many opportunities for error (Bandalos
& Boehm-Kaufman, 2009; Fabrigar et al., 1999; Hancock & Schoonen,
2015; Hurley et al., 1997; McCoach et al., 2013; Steiger, 2001; Stewart, 2001;
Tarka, 2018; Widaman, 2012). The users of CFA, in particular, are often
unaware of the limitations of that methodology (Brannick, 1995; DiStefano
& Hess, 2005; Hayduk, 2014; Kline, 2000, 2012; Ropovik, 2015; Schmitt
et al., 2018), possess misconceptions about its application (MacCallum &
Austin, 2000), and "tend to overstate both the strength and certainty of [its]
conclusions" (Tomarken & Waller, 2005, p. 48). It appears that many users
of CFA are unaware that the many possible choices for each decision lead
to a "garden of forking paths" (Gelman & Loken, 2014, p. 460), resulting
in findings that do not replicate. Hussey and Hughes (2020) referred to this
practice as v-hacking: "selectively choosing and reporting a combination of
metrics, including their implementations and cutoffs, and taking advantage
of other degrees of experimenter freedom so as to improve the apparent validity of measures" (p. 180).

Accordingly, the use of EFA may be preferable when: (a) "a strong empirical or conceptual foundation to guide the specification and evaluation of
factor model" is missing (Brown, 2013, p. 258); (b) the measured variables
are not multivariate normally distributed (Lei & Wu, 2012); or (c) the model
is incorrectly specified (Browne, 2001; Garrido et al., 2016; Kano, 1997;
Montoya & Edwards, 2021). Additionally, categorical measured variables,
especially test items, may be more productively analyzed with EFA (Ferrando
& Lorenzo-Seva, 2000; Schumacker & Lomax, 2004), especially when the
number of factors is in doubt because Garrido et al. (2016) demonstrated
that parallel analysis was more accurate than fit indices in determining the
number of factors. Likewise, EFA may be more appropriate when items fail
to meet the assumption of the independent clusters model (Marsh et al.,
2009; Molina et al., 2020). EFA may also be superior to the post hoc model
respecification of poor-fitting CFA models (Bandalos & Finney, 2019;
Browne, 2001; Flora & Flake, 2017; Gorsuch, 1997; Schmitt, 2011; Schmitt
et al., 2018; Warne & Burningham, 2019; Wegener & Fabrigar, 2000).

In contrast, CFA may be a more appropriate choice when: (a) strong conceptual and empirical foundations are available for specification of a model

(Orcan, 2018; Thompson, 2004); (b) the purpose is to test the invariance of a factor model across multiple groups (Osborne et al., 2007); or (c) direct estimation of a bifactor model is desired (Mansolf & Reise, 2015, 2016). Therefore, researchers should "not abandon EFA for the more recently developed confirmatory methods, but develop a heuristic strategy that builds on the comparative strengths of the two techniques" (Gerbing & Hamilton, 1996, p. 63).

Interestingly, similar issues have been explored in the past but within different contexts. Tukey (1980) asserted that "neither exploratory nor confirmatory is sufficient alone" (p. 23) and Box (1976) noted that science can only progress via "a motivated *iteration* between theory and practice" (p. 791). Thus, some facts lead to a tentative theory and then deductions from that tentative theory are found to be discrepant with other facts which, in turn, generates a modified theory. This abductive, deductive, and inductive cycle continues almost indefinitely as science progresses (Haig, 2018; Mulaik, 2018; Rummel, 1967). However, there are two impediments to this progress described by Box (1976) as *cookbookery* and *mathematistry*. "The symptoms of the former are a tendency to force all problems into the molds of one or two routine techniques" whereas mathematistry "is characterized by development of theory for theory's sake" (p. 797). Cookbookery can be recognized by routine use of a statistical method without consideration of its assumptions whereas mathematistry can be seen when researchers adopt inappropriate statistical procedures they do not understand. Consequently, the choice of factor analysis method must not degenerate into cookbookery and should not be determined solely by statistical complexity.

19 Practice Exercises

Exercise 1

To this point, readers have had an opportunity to conduct best-practice, evidence-based exploratory factor analysis (EFA) with both continuous and categorical data sets. Each EFA has followed a step-by-step process and each EFA has included Stata commands and results along with a discussion of the relevant principles. This first practice dataset is provided as an independent exercise. Readers are encouraged to analyze this dataset and check their results with the sample report provided at the end of this exercise.

Data

The Rmotivate.xlsx dataset can be downloaded as per the Data chapter and imported into the Stata environment as per the Importing and Saving Data chapter. Remember to check the imported data to ensure that its type and level of measurement are accurate. Alternatively, both Pearson and polychoric correlation matrices are provided in the RmotivateCorr.xlsx file.

Variables Included

The 20 reading motivation items (Figure 19.1) in this dataset are hypothesized to reflect two aspects of reading motivation: reading self-concept (odd items) and value of reading (even items). Each item offered four ordered response options.

Participants

The 20 reading motivation items were answered by 500 elementary school students equally distributed across grades 2 through 6.

Are the Data Appropriate?

Are data values accurate and plausible? Is there missing data and if so, how much and how was it handled? Are there outliers that bias the results? Are these

DOI: 10.4324/9781003149286-19

No	Description	No	Description
1.	My friends think I am	11.	I have trouble with reading
2.	Read a book	12.	Reading well is
3.	Reading skill	13.	I can answer teacher questions
4.	My friends think reading is	14.	I think reading is boring-interesting
5.	Can figure out unknown	15.	For me, reading is easy-hard
6.	Tell friends about books	16.	Will spend time reading when adult
7.	Understand what I read	17.	I understand reading assignments
8.	People who read are	18.	Would like more reading time
9.	As a reader, I am	19.	When reading aloud
10.	I think libraries are	20.	Books as presents

Figure 19.1 Description of the variables in the Rmotivate data

data normally distributed? If not, how nonnormal are they? Consequently, what correlation matrix should be employed for the EFA and why?

Is EFA Appropriate?

Why or why not? Be sure to report quantitative evidence to support that decision.

What Model of Factor Analysis Was Employed?

Choice between principal components analysis and common factor analysis. Rationale for that choice.

What Factor Extraction Method Was Used?

Specify and provide a rationale for that choice.

How Many Factors Were Retained?

What methods were selected to guide this decision? Rationale for choice of methods.

What Factor Rotation Was Applied?

Rationale for choice of orthogonal or oblique rotation and specific type of orthogonal or oblique rotation.

Interpretation of Results

Employ a model selection approach with a priori guidelines for interpretation.

Report Results

Sufficient detail to demonstrate that evidence-based decisions were made and results are professionally competent and reasonable. Follow the step-by-step checklist.

Compare your report to the sample report that follows.

Exercise 1 Report

Method

Participants were 500 elementary school students equally distributed across grades 2 through 6. Rouquette and Falissard (2011) simulated ordinal scale data and reported that scales with ten items and two factors required at least 350 participants to obtain stable and accurate factor analysis results. Consequently, the current sample size of 500 participants was judged to be adequate.

Given the uncertainty surrounding the structure of this reading motivation scale, exploratory factor analysis implemented in Version 16.1 of Stata was employed (Flora, 2018). Common factor analysis was selected over principal components analysis because the goal of this study was to identify the scale's latent structure (Widaman, 2018). Additionally, common factor analysis may produce more accurate estimates of population parameters than does principal components analysis (Widaman, 1993). Principal factor extraction was applied due to its relative tolerance of multivariate nonnormality and its superior recovery of weak factors (Briggs & MacCallum, 2003) and communalities were initially estimated by squared multiple correlations (Flora, 2018).

Lozano et al. (2008) conducted a statistical simulation study and found that four response categories, as offered by this reading motivation scale, were minimal to ensure adequate factorial validity with Pearson correlations. However, other methodologists have suggested the use of Pearson correlations only if there are at least five ordered categories and the variables are not severely nonnormal (DiStefano, 2002; Mueller & Hancock, 2019). For this purpose, univariate skew should not exceed 2.0 and univariate kurtosis should not exceed 7.0 (Curran et al., 1996). Given that the largest skew value was 2.1 and the largest kurtosis value was 3.8, Pearson correlations might be appropriate. Nevertheless, a sensitivity analysis will be performed with polychoric correlations to ensure that the results are robust.

Following the advice of Velicer et al. (2000), minimum average partials (MAP; Velicer, 1976) and parallel analysis (Horn, 1965), supplemented by a visual scree test (Cattell, 1966), were used to determine the number of factors to retain for rotation. For both theoretical and empirical reasons, it was assumed that factors would be correlated (Gorsuch, 1997; Meehl, 1990). Thus, a Promax rotation with a k value of 3 was selected (Gorsuch, 2003; Tataryn et al., 1999). To ensure both practical (10% variance explained) and statistical significance ($p < .01$), the threshold for salience was set at .32 (Norman & Streiner, 2014).

Some evidence favors overestimating rather than underestimating the number of factors (Wood et al., 1996); therefore, experts suggest that the highest to lowest number of factors be examined until the most interpretable solution is found (Fabrigar et al., 1999; Flora, 2018). Guidelines for model acceptability included: (a) at least three salient item loadings (pattern coefficients) are necessary to form a factor with the exclusion of complex loadings (Gorsuch, 1997); (b) an internal consistency reliability coefficient (alpha) of at least .80 for each factor because the intended use of this scale is for non-critical decisions about individual students (DeVellis, 2017); (c) no obvious symptoms of model misfit due to overfactoring or underfactoring; and (d) robustness of results across alternative extraction and rotation methods.

Results

There were no obvious illegal or out-of-bounds values and there was no missing data. A determinant value of .001 indicated that multicollinearity is not a serious issue. The Kaiser–Meyer–Olkin (KMO) measure of sampling adequacy (Kaiser, 1974) was acceptable (.93 for the total group of variables and .91 to .94 for each of the measured variables). Bartlett's test of sphericity (1950) statistically rejected the hypothesis that the correlation matrix was an identity matrix (chi-square of 3687.5 with 190 degrees of freedom at $p <$ .001). A visual scan of the correlation matrix revealed numerous coefficients $\geq$ .30 (Tabachnick & Fidell, 2019). Altogether, these measures indicate that factor analysis is appropriate.

Parallel analysis, MAP, and scree criteria were in agreement that two factors should be extracted for rotation and subsequent interpretation. However, models with three, two, and one factor(s) were sequentially evaluated for acceptability with the aforementioned guidelines.

The three-factor model explained 43% of the total variance before rotation, with the first, second, and third factors accounting for 31.4%, 9.3%, and 2.2% of the variance, respectively. The third factor was saliently loaded by five items, but two of those items were complex and that scale's alpha reliability was only .69, inadequate for most purposes. The root mean squared residual (RMSR) was .028 and there were no residual coefficients greater than .10. Thus, there was no indication that another factor should be extracted. However, the complex and unreliable structure of the third factor suggests that one less factor might be appropriate.

The two-factor model explained 40.5% of the total variance before rotation. Prior to rotation, the first factor (reading self-concept) accounted for 31.3% of the total variance while the second factor (value of reading) accounted for 9.2% of the total variance. This two-factor model was consistent with the theoretical underpinning of the scale with ten items loading saliently on a value of reading factor (.42 to .81) and ten items on a reading self-concept factor (.44 to .79) in a simple structure configuration. The interfactor correlation of .49 was low enough to pose no threat to discriminant validity (Brown, 2015) and the factors exhibited alpha reliability coefficients of .86, 95% CI [.85, .88] for the reading self-concept factor and .87, 95% CI [.85, .88] for the value of reading factor.

The RMSR for the two-factor model was .035 and there were only two residual coefficients greater than .10, suggesting that little residual variance remained after extracting two factors. Essentially, this model approached a close fit to the data (Maydeu-Olivares, 2017).

The one-factor model accounted for 30.8% of the total variance. All 20 items loaded saliently in the one-factor model but the RMSR reached an unacceptable level (.098) and 72 of the nonredundant residuals were > .10 (Brown, 2015). This is evidence of poor fit and suggests that two factors collapsed onto one (Fabrigar & Wegener, 2012). Thus, measures of model fit as well as theoretical convergence remove this model from consideration, leaving the two-factor model as the preferred solution.

Similar results were obtained when the extraction method was changed to maximum likelihood and the rotation method to oblimin. Given nonnormal categorical data with only four response options, a polychoric correlation matrix was also submitted for analysis and also identified two factors (Flora & Flake, 2017; Mueller & Hancock, 2019). Thus, these results were robust in demonstrating that this reading motivation scale is best described by two factors, reading self-concept and value of reading.

A Stata do-file to complete this EFA is provided in Figure 19.2.

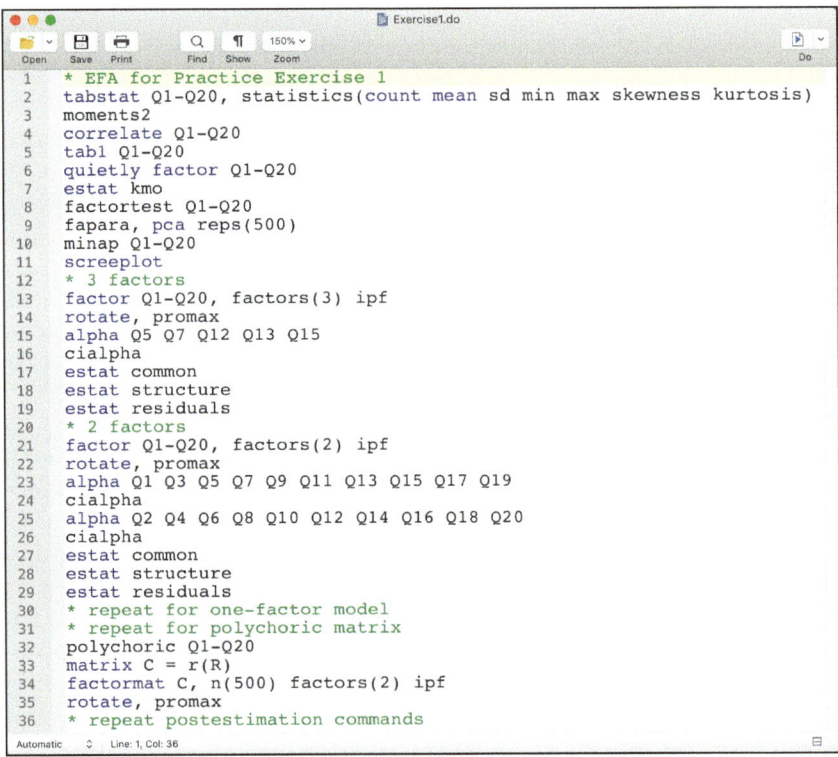

```
 1   * EFA for Practice Exercise 1
 2   tabstat Q1-Q20, statistics(count mean sd min max skewness kurtosis)
 3   moments2
 4   correlate Q1-Q20
 5   tabl Q1-Q20
 6   quietly factor Q1-Q20
 7   estat kmo
 8   factortest Q1-Q20
 9   fapara, pca reps(500)
10   minap Q1-Q20
11   screeplot
12   * 3 factors
13   factor Q1-Q20, factors(3) ipf
14   rotate, promax
15   alpha Q5 Q7 Q12 Q13 Q15
16   cialpha
17   estat common
18   estat structure
19   estat residuals
20   * 2 factors
21   factor Q1-Q20, factors(2) ipf
22   rotate, promax
23   alpha Q1 Q3 Q5 Q7 Q9 Q11 Q13 Q15 Q17 Q19
24   cialpha
25   alpha Q2 Q4 Q6 Q8 Q10 Q12 Q14 Q16 Q18 Q20
26   cialpha
27   estat common
28   estat structure
29   estat residuals
30   * repeat for one-factor model
31   * repeat for polychoric matrix
32   polychoric Q1-Q20
33   matrix C = r(R)
34   factormat C, n(500) factors(2) ipf
35   rotate, promax
36   * repeat postestimation commands
```

Figure 19.2 Do-file for EFA of Rmotivate data

Exercise 2

This fourth dataset (second practice exercise) is also provided for independent practice. Readers are encouraged to analyze this dataset while answering the questions found at the end of this chapter. No report, do-file, nor Stata results are provided with this practice exercise.

Data

This dataset is also in spreadsheet format and contains ten variables (Figure 19.3). Each variable is one item from a rating scale designed to tap the symptoms of attention-deficit hyperactivity disorder (ADHD) that was completed by 500 respondents. This data file labeled "adhd.xlsx" can be imported as was previously described in the Importing Raw Data section using the menu commands select ***File > Import > Excel spreadsheet***. Respondents reported the frequency of each item on a four-point Likert scale: 0 (*Never or Rarely*), 1 (*Sometimes*), 2 (*Often*), and 3 (*Very Often*). Thus, the data are ordered and not continuous.

These ten items are hypothesized to reflect two behavioral aspects of ADHD: attention problems and over-activity/impulsivity problems. It was assumed that the first five items tap the attention problems dimension while the final five items tap the over-activity/impulsivity dimension of ADHD.

Item	Description
instruct	Follow instructions
effort	Sustain mental effort
organize	Organization problems
forget	Forgetful
attention	Sustain attention
go	Constantly on the go
talks	Talk excessively
fidgets	Fidget
turns	Difficulty waiting turn
runs	Runs about

Figure 19.3 Description of the variables in the ADHD dataset

Scales with more items have typically found internal consistency reliability coefficients of around .85 to .90 for these factors (Nichols et al., 2017).

Questions to Answer

1. Are the data appropriate for EFA? Address linearity, outliers, normality, missing data.
2. Is EFA appropriate? Verify with KMO sampling adequacy and Bartlett's test of sphericity.
3. What correlation matrix should be used? See Figure 2.7 for both Pearson and polychoric matrices.
4. What factor analysis model is appropriate? Justify choice of principal components or common factor models.
5. What factor extraction method is appropriate? Detail the method of factor extraction and correlation matrix employed.
6. How many factors to retain? Describe and justify the selection criteria.
7. What factor rotation method is appropriate? Justify rotation method (orthogonal or oblique) and type (varimax, promax, oblimin, etc.).
8. Interpret each model using a priori guidelines for acceptability.
9. Report results of each decision step.

References and Resources

Acock, A. C. (2018). *A gentle introduction to stata* (5th ed.). Stata Press.

Aiken, L. S., West, S. G., & Millsap, R. E. (2008). Doctoral training in statistics, measurement, and methodology in psychology. *American Psychologist*, *63*(1), 32–50. https://doi.org/10.1037/0003-066X.63.1.32

Anderson, S. F. (2020). Misinterpreting *p*: The discrepancy between *p* values and the probability the null hypothesis is true, the influence of multiple testing, and implications for the replication crisis. *Psychological Methods*, *25*(5), 596–609. https://doi.org/10.1037/met0000248

Anscombe, F. J. (1973). Graphs in statistical analysis. *The American Statistician*, *27*(1), 17–21. https://doi.org/10.2307/2682899

Armstrong, J. S., & Soelberg, P. (1968). On the interpretation of factor analysis. *Psychological Bulletin*, *70*(5), 361–364. https://doi.org/10.1037/h0026434

Auerswald, M., & Moshagen, M. (2019). How to determine the number of factors to retain in exploratory factor analysis: A comparison of extraction methods under realistic conditions. *Psychological Methods*, *24*(4), 468–491. http://dx.doi.org/10.1037/met0000200

Baldwin, S. A. (2019). *Psychological statistics and psychometrics using Stata*. Stata Press.

Bandalos, D. L. (2018). *Measurement theory and applications in the social sciences*. Guilford.

Bandalos, D. L., & Boehm-Kaufman, M. R. (2009). Four common misconceptions in exploratory factor analysis. In C. E. Lance & R. J. Vandenberg (Eds.), *Statistical and methodological myths and urban legends: Doctrine, verity and fable in the organizational and social sciences* (pp. 61–87). Routledge.

Bandalos, D. L., & Finney, S. J. (2019). Factor analysis: Exploratory and confirmatory. In G. R. Hancock, L. M. Stapleton, & R. O. Mueller (Eds.), *The reviewer's guide to quantitative methods in the social sciences* (2nd ed., pp. 98–122). Routledge.

Bandalos, D. L., & Gerstner, J. J. (2016). Using factor analysis in test construction. In K. Schweizer & C. DiStefano (Eds.), *Principles and methods of test construction: Standards and recent advances* (pp. 26–51). Hogrefe.

Baraldi, A. N., & Enders, C. K. (2013). Missing data methods. In T. D. Little (Ed.), *Oxford handbook of quantitative methods: Statistical analyses* (Vol. 2, pp. 635–664). Oxford University Press.

Barendse, M. T., Oort, F. J., & Timmerman, M. E. (2015). Using exploratory factor analysis to determine the dimensionality of discrete responses. *Structural Equation Modeling*, *22*(1), 87–101. https://doi.org/10.1080/10705511.2014.934850

Barrett, P. (2007). Structural equation modelling: Adjudging model fit. *Personality and Individual Differences*, *42*(5), 815–824. https://doi.org/10.1016/j.paid.2006.09.018

Barrett, P. T., & Kline, P. (1982). Factor extraction: An examination of three methods. *Personality Study and Group Behaviour, 3*(1), 84–98.

Bartholomew, D. J. (1995). Spearman and the origin and development of factor analysis. *British Journal of Mathematical and Statistical Psychology, 48*(2), 211–220. https://doi.org/10.1111/j.2044-8317.1995.tb01060.x

Bartlett, M. S. (1937). The statistical conception of mental factors. *British Journal of Psychology, 28*(1), 97–104. https://doi.org/10.1111/j.2044-8295.1937.tb00863.x

Bartlett, M. S. (1950). Tests of significance in factor analysis. *British Journal of Psychology, 3*(2), 77–85. https://doi.org/10.1111/j.2044-8317.1950.tb00285.x

Basto, M., & Pereira, J. M. (2012). An SPSS R-menu for ordinal factor analysis. *Journal of Statistical Software, 46*(4), 1-29.

Baum, C. F. (2016). *An introduction to Stata programming* (2nd ed.). Stata Press.

Beaujean, A. A. (2013). Factor analysis using R. *Practical Assessment, Research & Evaluation, 18*(4), 1–11. https://doi.org/10.7275/z8wr-4j42

Beavers, A. S., Lounsbury, J. W., Richards, J. K., Huck, S. W., Skolits, G. J., & Esquivel, S. L. (2012). Practical considerations for using exploratory factor analysis in educational research. *Practical Assessment, Research & Evaluation, 18*(6), 1–13. https://doi.org/10.7275/qv2q-rk76

Bedeian, A. G., Sturman, M. C., & Streiner, D. L. (2009). Decimal dust, significant digits, and the search for stars. *Organizaitonal Research Methods, 12*(4), 687–694. https://doi.org/10.1177/1094428108321153

Benjamin, D. J., & Berger, J. O. (2019). Three recommendations for improving the use of *p*-values. *The American Statistician, 73*(S1), 186–191. https://doi.org/10.1080/00031305.2018.1543135

Benson, J. (1998). Developing a strong program of construct validation: A test anxiety example. *Educational Measurement: Issues and Practice, 17*(1), 10–22. https://doi.org/10.1111/j.1745-3992.1998.tb00616.x

Benson, J., & Nasser, F. (1998). On the use of factor analysis as a research tool. *Journal of Vocational Education Research, 23*(1), 13–33.

Bentler, P. M. (2005). *EQS 6 structural equations program manual.* Multivariate Software.

Berk, R. (2011). Evidence-based versus junk-based evaluation research: Some lessons from 35 years of the Evaluation Review. *Evaluation Review, 35*(3), 191–203. https://doi.org/10.1177/0193841X11419281

Bernstein, I. H., & Teng, G. (1989). Factoring items and factoring scales are different: Spurious evidence for multidimensionality due to item categorization. *Psychological Bulletin, 105*(3), 467–477. https://doi.org/10.1037/0033-2909.105.3.467

Bishara, A. J., & Hittner, J. B. (2015). Reducing bias and error in the correlation coefficient due to nonnormality. *Educational and Psychological Measurement, 75*(5), 785–804. https://doi.org/10.1177/0013164414557639

Bollen, K. A. (2002). Latent variables in psychology and the social sciences. *Annual Review of Psychology, 53*(1), 605–634. https://doi.org/10.1146/annurev.psych.53.100901.135239

Bollen, K. A., & Barb, K. H. (1981). Pearson's R and coarsely categorized measures. *American Sociological Review, 46*(2), 232–239. https://doi.org/10.2307/2094981

Box, G. E. P. (1976). Science and statistics. *Journal of the American Statistical Association, 71*(356), 791–799. http://doi.org/10.1080/01621459.1976.10480949

Brannick, M. T. (1995). Critical comments on applying covariance structure modeling. *Journal of Organizational Behavior, 16*(3), 201–213. https://doi.org/10.1002/job.4030160303

Breckler, S. J. (1990). Applications of covariance structure modeling in psychology: Cause for concern? *Psychological Bulletin*, *107*(2), 260–273. https://doi.org/10.1037/0033-2909.107.2.260

Briggs, S. R., & Cheek, J. M. (1986). The role of factor analysis in the development and evaluation of personality scales. *Journal of Personality*, *54*(1), 106–148. https://doi.org/10.1111/j.1467-6494.1986.tb00391.x

Briggs, N. E., & MacCallum, R. C. (2003). Recovery of weak common factors by maximum likelihood and ordinary least squares estimation. *Multivariate Behavioral Research*, *38*(1), 25–56. https://doi.org/10.1207/S15327906MBR3801_2

Brown, T. A. (2013). Latent variable measurement models. In T. D. Little (Ed.), *Oxford handbook of quantitative methods: Statistical analysis* (Vol. 2, pp. 257–280). Oxford University Press.

Brown, T. A. (2015). *Confirmatory factor analysis for applied research* (2nd ed.). Guilford.

Browne, M. W. (2001). An overview of analytic rotation in exploratory factor analysis. *Multivariate Behavioral Research*, *36*(1), 111–150. https://doi.org/10.1207/S15327906MBR3601_05

Browne, M. W., & Cudeck, R. (1993). Alternative ways of assessing model fit. In K. A. Bollen & J. S. Long (Eds.), *Testing structural equation models* (pp. 136–162). Sage.

Brunner, M., Nagy, G., & Wilhelm, O. (2012). A tutorial on hierarchically structured constructs. *Journal of Personality*, *80*(4), 796–846. https://doi.org/10.1111/j.1467-6494.2011.00749.x

Budaev, S. V. (2010). Using principal components and factor analysis in animal behaviour research: Caveats and guidelines. *Ethology*, *116*(5), 472–480. https://doi.org/10.1111/j.1439-0310.2010.01758.x

Buja, A., & Eyuboglu, N. (1992). Remarks on parallel analysis. *Multivariate Behavioral Research*, *27*(4), 509–540. https://doi.org/10.1207/s15327906mbr2704_2

Burnham, K. P., & Anderson, D. R. (2004). Multimodel inference: Understanding AIC and BIC in model selection. *Sociological Methods & Research*, *33*(2), 261–304. https://doi.org/10.1177/0049124104268644

Burt, C. (1940). *The factors of the mind: An introduction to factor-analysis in psychology*. University of London Press.

Cabrera-Nguyen, P. (2010). Author guidelines for reporting scale development and validation results in the *Journal of the Society for Social Work and Research*. *Journal of the Society for Social Work and Research*, *1*(2), 99–103. https://doi.org/10.5243/jsswr.2010.8

Cain, M. K., Zhang, Z., & Yuan, K.-H. (2017). Univariate and multivariate skewness and kurtosis for measuring nonnormality: Prevalence, influence, and estimation. *Behavior Research Methods*, *49*(5), 1716–1735. https://doi.org/10.3758/s13428-016-0814-1

Canivez, G. L., & Watkins, M. W. (2010). Exploratory and higher-order factor analyses of the Wechsler Adult Intelligence Scale-Fourth Edition (WAIS-IV) adolescent subsample. *School Psychology Quarterly*, *25*(4), 223–235. https://doi.org/10.1037/a0022046

Canivez, G. L., Watkins, M. W., & Dombrowski, S. C. (2016). Factor structure of the Wechsler Intelligence Scale for Children-Fifth Edition: Exploratory factor analysis with the 16 primary and secondary subtests. *Psychological Assessment*, *28*(8), 975–986. https://doi.org/10.1037/pas0000238

Caron, P.-O. (2019). Minimum average partial correlation and parallel analysis: The influence of oblique structures. *Communications in Statistics: Simulation and Computation*, *48*(7), 2110–2117. https://doi.org/10.1080/03610918.2018.1433843

Carretta, T. R., & Ree, J. J. (2001). Pitfalls of ability research. *International Journal of Selection and Assessment*, *9*(4), 325–335. https://doi.org/10.1111/1468-2389.00184

Carroll, J. B. (1961). The nature of the data, or how to choose a correlation coefficient. *Psychometrika, 26*(4), 347–372. https://doi.org/10.1007/BF02289768

Carroll, J. B. (1978). How shall we study individual differences in cognitive abilities?– Methodological and theoretical perspectives. *Intelligence, 2*(2), 87–115. https://doi.org/10.1016/0160-2896(78)90002-8

Carroll, J. B. (1983). Studying individual differences in cognitive abilities: Through and beyond factor analysis. In R. F. Dillon & R. R. Schmeck (Eds.), *Individual differences in cognition* (pp. 1–33). Academic Press.

Carroll, J. B. (1985). Exploratory factor analysis: A tutorial. In D. K. Detterman (Ed.), *Current topics in human intelligence* (pp. 25–58). Ablex Publishing Company.

Carroll, J. B. (1993). *Human cognitive abilities: A survey of factor-analytic studies*. Cambridge University Press.

Carroll, J. B. (1995a). On methodology in the study of cognitive abilities. *Multivariate Behavioral Research, 30*(3), 429–452. https://doi.org/10.1207/s15327906mbr3003_6

Carroll, J. B. (1995b). Reflections on Stephen Jay Gould's *the Mismeasure of Man* (1981): A retrospective review. *Intelligence, 21*(2), 121–134. https://doi.org/10.1016/0160-2896(95)90022-5

Cattell, R. B. (1946). *The description and measurement of personality*. World Book.

Cattell, R. B. (1952). *Factor analysis: An introduction and manual for the psychologist and social scientist*. Greenwood Press.

Cattell, R. B. (1966). The scree test for the number of factors. *Multivariate Behavioral Research, 1*(2), 245–276. https://doi.org/10.1207/s15327906mbr0102_10

Cattell, R. B. (1978). *The scientific use of factor analysis in behavioral and life sciences*. Plenum Press.

Chen, F. F., West, S. G., & Sousa, K. H. (2006). A comparison of bifactor and second-order models of quality of life. *Multivariate Behavioral Research, 41*(2), 189–225.

Chen, S.-F., Wang, S., & Chen, C.-Y. (2012). A simulation study using EFA and CFA programs based on the impact of missing data on test dimensionality. *Expert Systems with Applications, 39*(4), 4026–4031. https://doi.org/10.1016/j.eswa.2011.09.085

Child, D. (2006). *The essentials of factor analysis* (3rd ed.). Continuum.

Cho, S.-J., Li, F., & Bandalos, D. (2009). Accuracy of the parallel analysis procedure with polychoric correlations. *Educational and Psychological Measurement, 69*(5), 748–759. https://doi.org/10.1177/0013164409332229

Choi, J., Peters, M., & Mueller, R. O. (2010). Correlational analysis of ordinal data: From Pearson's r to Bayesian polychoric correlation. *Asia Pacific Education Review, 11*(4), 459–466. https://doi.org/10.1007/s12564-010-9096-y

Clark, D. A., & Bowles, R. P. (2018). Model fit and item factor analysis: Overfactoring, underfactoring, and a program to guide interpretation. *Multivariate Behavioral Research, 53*(4), 544–558. https://doi.org/10.1080/00273171.2018.1461058

Cliff, N. (1983). Some cautions concerning the application of causal modeling methods. *Multivariate Behavioral Research, 18*(1), 115–126. https://doi.org/10.1207/s15327906mbr1801_7

Clifton, J. D. W. (2020). Managing validity versus reliability trade-offs in scale-building decisions. *Psychological Methods, 25*(3), 259–270. https://doi.org/10.1037/met0000236

Comrey, A. L. (1988). Factor-analytic methods of scale development in personality and clinical psychology. *Journal of Consulting and Clinical Psychology, 56*(5), 754–761. https://doi.org/10.1037//0022-006x.56.5.754

Comrey, A. L., & Lee, H. B. (1992). *A first course in factor analysis* (2nd ed.). Erlbaum.

Conway, J. M., & Huffcutt, A. I. (2003). A review and evaluation of exploratory factor analysis practices in organizational research. *Organizational Research Methods, 6*(2), 147–168. https://doi.org/10.1177/1094428103251541

Cooper, C. (2019). *Psychological testing: Theory and practice.* Routledge.

Costello, A. B., & Osborne, J. W. (2005). Best practices in exploratory factor analysis: Four recommendations for getting the most from your analysis. *Practical Assessment, Research & Evaluation, 10*(7), 1–9. https://doi.org/10.7275/jyj1-4868

Crawford, A. V., Green, S. B., Levy, R., Lo, W.-J., Scott, L., Svetina, D., & Thompson, M. S. (2010). Evaluation of parallel analysis methods for determining the number of factors. *Educational and Psychological Measurement, 70*(6), 885–901. https://doi.org/10.1177/0013164410379332

Cribbie, R. A. (2000). Evaluating the importance of individual parameters in structural equation modeling: The need for type I error control. *Personality and Individual Differences, 29*(3), 567–577. https://doi.org/10.1016/S0191-8869(99)00219-6

Cronbach, L. J. (1951). Coefficient alpha and the internal structure of tests. *Psychometrika, 16*(3), 297–334. https://doi.org/10.1007/BF02310555

Cudeck, R. (2000). Exploratory factor analysis. In H. E. A. Tinsley & S. D. Brown (Eds.), *Handbook of applied multivariate statistics and mathematical modeling* (pp. 265–296). Academic Press.

Curran, P. J. (2016). Methods for the detection of carelessly invalid responses in survey data. *Journal of Experimental Social Psychology, 66*, 4–19. https://doi.org/10.1016/j.jesp.2015.07.006

Curran, P. J., West, S. G., & Finch, J. F. (1996). The robustness of test statistics to nonnormality and specification error in confirmatory factor analysis. *Psychological Methods, 1*(1), 16–29. https://doi.org/10.1037/1082-989X.1.1.16

DeCarlo, L. T. (1997). On the meaning and use of kurtosis. *Psychological Methods, 2*(3), 292–307. https://doi.org/10.1037/1082-989X.2.3.292

DeSimone, J. A., Harms, P. D., & DeSimone, A. J. (2015). Best practice recommendations for data screening. *Journal of Organizational Behavior, 36*(2), 171–181. https://doi.org/10.1002/job.1962

DeVellis, R. F. (2017). *Scale development: Theory and applications* (4th ed.). Sage.

de Winter, J. C. F., & Dodou, D. (2012). Factor recovery by principal axis factoring and maximum likelihood factor analysis as a function of factor pattern and sample size. *Journal of Applied Statistics, 39*(4), 695–710. https://doi.org/10.1080/02664763.2011.610445

de Winter, J. C. F., Dodou, D., & Wieringa, P. A. (2009). Exploratory factor analysis with small sample sizes. *Multivariate Behavioral Research, 44*(2), 147–181. https://doi.org/10.1080/00273170902794206

de Winter, J. C. F., Gosling, S. D., & Potter, J. (2016). Comparing the Pearson and Spearman correlation coefficients across distributions and sample sizes: A tutorial using simulations and empirical data. *Psychological Methods, 21*(3), 273–290. https://doi.org/10.1037/met0000079

Digman, J. M. (1990). Personality structure: Emergence of the five-factor model. *Annual Review of Psychology, 41*, 417–440. https://doi.org/10.1146/annurev.ps.41.020190.002221

Dinno, A. (2009). Exploring the sensitivity of Horn's parallel analysis to the distributional form of random data. *Multivariate Behavioral Research, 44*(3), 362–388. https://doi.org/10.1080/00273170902938969

DiStefano, C. (2002). The impact of categorization with confirmatory factor analysis. *Structural Equation Modeling, 9*(3), 327–346. https://doi.org/10.1207/S15328007SEM0903_2

DiStefano, C. (2016). Examining fit with structural equation models. In K. Schweizer & C. DiStefano (Eds.), *Principles and methods of test construction: Standards and recent advances* (pp. 166–193). Hogrefe.

DiStefano, C., & Hess, B. (2005). Using confirmatory factor analysis for construct validation: An empirical review. *Journal of Psychoeducational Assessment, 23*(3), 225–241. https://doi.org/10.1177/073428290502300303

DiStefano, C., Shi, D., & Morgan, G. B. (2021). Collapsing categories is often more advantageous than modeling sparse data: Investigations in the CFA framework. *Structural Equation Modeling, 28*(7), 237-249. https://doi.org/10.1080/10705511.2020.1803073

DiStefano, C., Zhu, M., & Mindrila, D. (2009). Understanding and using factor scores: Considerations for the applied researcher. *Practical Assessment, Research & Evaluation, 14*(20), 1–11. https://doi.org/10.7275/da8t-4g52

Dombrowski, S. C., Beaujean, A. A., McGill, R. J., Benson, N. F., & Schneider, W. J. (2019). Using exploratory bifactor analysis to understand the latent structure of multidimensional psychological measures: An example featuring the WISC-V. *Structural Equation Modeling, 26*(6), 847–860. https://doi.org/10.1080/10705511.2019.1622421

Dombrowski, S. C., Canivez, G. L., & Watkins, M. W. (2018). Factor structure of the 10 WISC-V primary subtests across four standardization age groups. *Contemporary School Psychology, 22*(1), 90–104. https://doi.org/10.1007/s40688-017-0125-2

Dombrowski, S. C., McGill, R. J., Canivez, G. L., Watkins, M. W., & Beaujean, A. A. (2021). Factor analysis and variance partitioning in intelligence test research: Clarifying misconceptions. *Journal of Psychoeducational Assessment, 39*(2), 28–38. https://doi.org/10.1177/0734282920961952

Dunn, A. M., Heggestad, E. D., Shanock, L. R., & Theilgard, N. (2018). Intra-individual response variability as an indicator of insufficient effort responding: Comparison to other indicators and relationships with individual differences. *Journal of Business and Psychology, 33*(1), 105–121. https://doi.org/10.1007/s10869-016-9479-0

Dziuban, C. D., & Shirkey, E. S. (1974). When is a correlation matrix appropriate for factor analysis? Some decision rules. *Psychological Bulletin, 81*(6), 358–361. https://doi.org/10.1037/h0036316

Edwards, J. R., & Bagozzi, R. P. (2000). On the nature and direction of relationships between constructs and measures. *Psychological Methods, 5*(2), 155–174. https://doi.org/10.1037/1082-989x.5.2.155

Enders, C. K. (2017). Multiple imputation as a flexible tool for missing data handling in clinical research. *Behaviour Research and Therapy, 98*, 4–18. https://doi.org/10.1016/j.brat.2016.11.008

Epskamp, S., Maris, G., Waldorp, L. J., & Borsboom, D. (2018). In P. Iwing, T. Booth, & D. J. Hughes (Eds.), *The Wiley handbook of psychometric testing: A multidisciplinary reference on survey, scale and test development* (pp. 953–986). Wiley.

Fabrigar, L. R., & Wegener, D. T. (2012). *Exploratory factor analysis.* Oxford University Press.

Fabrigar, L. R., Wegener, D. T., MacCallum, R. C., & Strahan, E. J. (1999). Evaluating the use of exploratory factor analysis in psychological research. *Psychological Methods, 4*(3), 272–299. https://doi.org/10.1037/1082-989X.4.3.272

Fava, J. L., & Velicer, W. F. (1992). The effects of overextraction on factor and component analysis. *Multivariate Behavioral Research, 27*(3), 387–415. https://doi.org/10.1207/s15327906mbr2703_5

Fava, J. L., & Velicer, W. F. (1996). The effects of underextraction in factor and component analysis. *Educational and Psychological Measurement, 56*(6), 907–929. https://doi.org/10.1177/0013164496056006001

Feldt, L. S., & Brennan, R. L. (1993). Reliability. In R. L. Linn (Ed.), *Educational measurement* (3rd ed., pp. 105–146). Oryx Press.

Fernstad, S. J. (2019). To identify what is not there: A definition of missingness patterns and evaluation of missing value visualization. *Information Visualization, 18*(2), 230–250. https://doi.org/10.1177/1473871618785387

Ferrando, P. J., & Lorenzo-Seva, U. (2000). Unrestricted versus restricted factor analysis of multidimensional test items: Some aspects of the problem and some suggestions. *Psicológica, 21*(3), 301–323.

Ferrando, P. J., & Lorenzo-Seva, U. (2018). Assessing the quality and appropriateness of factor solutions and factor score estimates in exploratory item factor analysis. *Educational and Psychological Measurement, 78*(5), 762–780. https://doi.org/10.1177/0013164417719308

Ferrando, P. J., Navarro-González, D., & Lorenzo-Seva, U. (2019). Assessing the quality and effectiveness of the factor score estimates in psychometric factor-analytic applications. *Methodology, 15*(3), 119–127. https://doi.org/10.1027/1614-2241/a000170

Feynman, R. P. (1974). Cargo cult science. *Engineering and Science, 37*(7), 10–13. http://calteches.library.caltech.edu/51/1/ES37.7.1974.pdf

Field, A., Miles, J., & Field, Z. (2012). *Discovering statistics using R*. Sage.

Finch, H. (2006). Comparison of the performance of varimax and promax rotations: Factor structure recovery for dichotomous items. *Journal of Educational Measurement, 43*(1), 39–52. https://doi.org/10.1111/j.1745-3984.2006.00003.x

Finch, W. H. (2013). Exploratory factor analysis. In T. Teo (Ed.), *Handbook of quantitative methods for educational research* (pp. 167–186). Sense Publishers.

Finch, W. H. (2020a). *Exploratory factor analysis*. Sage.

Finch, W. H. (2020b). Using fit statistic differences to determine the optimal number of factors to retain in an exploratory factor analysis. *Educational and Psychological Measurement, 80*(2), 217–241. https://doi.org/10.1177/0013164419865769

Finney, S. J., & DiStefano, C. (2013). Nonnormal and categorical data in structural equation modeling. In G. R. Hancock & R. O. Mueller (Eds.), *Structural equation modeling: A second course* (2nd ed., pp. 439–492). Information Age Publishing.

Flake, J. K., & Fried, E. I. (2020). Measurement schmeasurement: Questionable measurement practices and how to avoid them. *Advances in Methods and Practices in Psychological Science, 3*(4), 456–465. https://doi.org/10.1177/2515245920952393

Flora, D. B. (2018). *Statistical methods for the social and behavioural sciences: A model-based approach*. Sage.

Flora, D. B. (2020). Your coefficient alpha is probably wrong, but which coefficient omega is right? A tutorial on using R to obtain better reliability estimates. *Advances in Methods and Practices in Psychological Science, 3*(4), 484–501. https://doi.org/10.1177/2515245920951747

Flora, D. B., & Curran, P. J. (2004). An empirical evaluation of alternative methods of estimation for confirmatory factor analysis with ordinal data. *Psychological Methods, 9*(4), 466–491. https://doi.org/10.1037/1082-989X.9.4.466

Flora, D. B., & Flake, J. K. (2017). The purpose and practice of exploratory and confirmatory factor analysis in psychological research: Decisions for scale development and validation. *Canadian Journal of Behavioural Science, 49*(2), 78–88. https://doi.org/10.1037/cbs0000069

Flora, D. B., LaBrish, C., & Chalmers, R. P. (2012). Old and new ideas for data screening and assumption testing for exploratory and confirmatory factor analysis. *Frontiers in Psychology*, *3*(55), 1–21. https://doi.org/10.3389/fpsyg.2012.00055

Floyd, F. J., & Widaman, K. F. (1995). Factor analysis in the development and refinement of clinical assessment instruments. *Psychological Assessment*, *7*(3), 286–299. https://doi.org/10.1037/1040-3590.7.3.286

Ford, J. K., MacCallum, R. C., & Tait, M. (1986). The application of exploratory factor analysis in applied psychology: A critical review and analysis. *Personnel Psychology*, *39*(2), 291–314. https://doi.org/10.1111/j.1744-6570.1986.tb00583.x

Freedman, D.A. (1987). A rejoinder on models, metaphors, and fables. *Journal of Educational Statistics*, *12*(2), 206–223. https://doi.org/10.3102/10769986012002206

French, J. W., Tucker, L. R., Newman, S. H., & Bobbitt, J. M. (1952). A factor analysis of aptitude and achievement entrance tests and course grades at the United States Coast Guard Academy. *Journal of Educational Psychology*, *43*(2), 65–80. https://doi.org/10.1037/h0054549

Garcia-Garzon, E., Abad, F. J., & Garriso, L. E. (2021). On omega hierarchical estimation: A comparison of exploratory bi-factor analysis algorithms. *Multivariate Behavioral Research*, *56*(1), 101–119. https://doi.org/10.1080/00273171.2020.1736977

Garrido, L. E., Abad, F. J., & Ponsoda, V. (2011). Performance of Velicer's minimum average partial factor retention method with categorical variables. *Educational and Psychological Measurement*, *71*(3), 551–570. https://doi.org/10.1177/0013164410389489

Garrido, L. E., Abad, F. J., & Ponsoda, V. (2013). A new look at Horn's parallel analysis with ordinal variables. *Psychological Methods*, *18*(4), 454–474. https://doi.org/10.1037/a0030005

Garrido, L. E., Abad, F. J., & Ponsoda, V. (2016). Are fit indices really fit to estimate the number of factors with categorical variables? Some cautionary findings via Monte Carlo simulation. *Psychological Methods*, *21*(1), 93–111. http://dx.doi.org/10.1037/met0000064

Garson, G. D. (2013). *Factor analysis*. Statistical Publishing Associates.

Gaskin, C. J., & Happell, B. (2014). Exploratory factor analysis: A review of recent evidence, an assessment of current practice, and recommendations for future use. *International Journal of Nursing Studies*, *51*(3), 511–521. https://doi.org/10.1016/j.ijnurstu.2013.10.005

Gelman, A., & Loken, E. (2014). The statistical crisis in science. *American Scientist*, *102*(6), 460–465. https://doi.org/10.1511/2014.111.460

Gerbing, D. W., & Hamilton, J. G. (1996). Viability of exploratory factor analysis as a precursor to confirmatory factor analysis. *Structural Equation Modeling*, *3*(1), 62–72. https://doi.org/10.1080/10705519609540030

Gibson, T. O., Morrow, J. A., & Rocconi, L. M. (2020). A modernized heuristic approach to robust exploratory factor analysis. *Quantitative Methods in Psychology*, *16*(4), 295–307. https://doi.org/10.20982/tqmp.16.4.p295

Gignac, G. E. (2007). Multi-factor modeling in individual differences research: Some recommendations and suggestions. *Personality and Individual Differences*, *42*(1), 37–48. https://doi.org/10.1016/j.paid.2006.06.019

Gignac, G. E. (2008). Higher-order models versus direct hierarchical models: g as superordinate or breadth factor? *Psychology Science Quarterly*, *50*(1), 21–43.

Gilman, R., Laughlin, J. E., & Huebner, E. S. (1999). Validation of the Self-Description Questionnaire-II with an American sample. *School Psychology International*, *20*(3), 300–307. https://doi.org/10.1177/0143034399203005

Giordano, C., & Waller, N. G. (2020). Recovering bifactor models: A comparison of seven methods. *Psychological Methods*, *25*(2), 143–156. https://doi.org/10.1037/met0000227

Glorfeld, L. W. (1995). An improvement on Horn's parallel analysis methodology for selecting the correct number of factors to retain. *Educational and Psychological Measurement*, *55*(3), 377–393. https://doi.org/10.1177/0013164495055003002

Goldberg, L. R., & Velicer, W. F. (2006). Principles of exploratory factor analysis. In S. Strack (Ed.), *Differentiating normal and abnormal personality* (2nd ed., pp. 209–237). Springer.

Golino, H. F., & Epskamp, S. (2017). Exploratory graph analysis: A new approach for estimating the number of dimensions in psychological research. *PlosOne*, *12*(6), 1–26. https://doi.org/10.1371/journal.pone.0174035

Golino, H., Shi, D., Christensen, A. P., Garrido, L. E., Nieto, M. D., Sadana, R., & Martinez-Molina, A. (2020). Investigating the performance of exploratory graph analysis and traditional techniques to identify the number of latent factors: A simulation and tutorial. *Psychological Methods*, *25*(3), 292–320. https://doi.org/10.1037/met0000255

Goodwin, L. D. (1999). The role of factor analysis in the estimation of construct validity. *Measurement in Physical Education and Exercise Science*, *3*(2), 85–100. https://doi.org/10.1207/s15327841mpee0302_2

Goodwin, L. D., & Leech, N. L. (2006). Understanding correlation: Factors that affect the size of r. *Journal of Experimental Education*, *74*(3), 251–266. https://doi.org/10.3200/JEXE.74.3.249-266

Goretzko, D., Pham, T. T. H., & Bühner, M. (2019). Exploratory factor analysis: Current use, methodological developments and recommendations for good practice. *Current Psychology*. https://doi.org/10.1007/s12144-019-00300-2

Gorsuch, R. L. (1983). *Factor analysis* (2nd ed.). Erlbaum.

Gorsuch, R. L. (1988). Exploratory factor analysis. In J. R. Nesselroade & R. B. Cattell (Eds.), *Handbook of multivariate experimental psychology* (2nd ed.). Plenum Press.

Gorsuch, R. L. (1990). Common factor analysis versus component analysis: Some well and little known facts. *Multivariate Behavioral Research*, *25*(1), 33–39. https://doi.org/10.1207/s15327906mbr2501_3

Gorsuch, R. L. (1997). Exploratory factor analysis: Its role in item analysis. *Journal of Personality Assessment*, *68*(3), 532–560. https://doi.org/10.1207/s15327752jpa6803_5

Gorsuch, R. L. (2003). Factor analysis. In J. A. Schinka & W. F. Velicer (Eds.), *Handbook of psychology: Research methods in psychology* (Vol. 2, pp. 143–164). Wiley.

Graham, J. M., Guthrie, A. C., & Thompson, B. (2003). Consequences of not interpreting structure coefficients in published CFA research: A reminder. *Structural Equation Modeling*, *10*(1), 142–153. https://doi.org/10.1207/S15328007SEM1001_7

Greenwald, A. G., Pratkanis, A. R., Leippe, M. R., & Baumgardner, M. H. (1986). Under what conditions does theory obstruct research progress? *Psychological Review*, *93*(2), 216–229. https://doi.org/10.1037/0033-295X.93.2.216

Greer, T., Dunlap, W. P., Hunter, S. T., & Berman, M. E. (2006). Skew and internal consistency. *Journal of Applied Psychology*, *91*(6), 1351–1358. https://doi.org/10.1037/0021-9010.91.6.1351

Greiff, S., & Heene, M. (2017). Why psychological assessment needs to start worrying about model fit. *European Journal of Psychological Assessment, 33*(5), 313–317. https://doi.org/10.1027/1015-5759/a000450

Grice, J. W. (2001). Computing and evaluating factor scores. *Psychological Methods, 6*(4), 430–450. https://doi.org/10.1037/1082-989X.6.4.430

Grieder, S., & Steiner, M. D. (2020). Algorithmic jingle jungle: A comparison of implementations of principal axis factoring and promax rotation in R and SPSS. *PsyArXiv.* https://psyarxiv.com/7hwrm

Guadagnoli, E., & Velicer, W. F. (1988). Relation of sample size to the stability of component patterns. *Psychological Bulletin, 103*(2), 265–275. https://doi.org/10.1037/0033-2909.103.2.265

Hägglund, G. (2001). Milestones in the history of factor analysis. In R. Cukeck, S. du Toit, & D. Sörbom (Eds.), *Structural equation modeling: Present and future* (pp. 11–38). Scientific Software International.

Hahs-Vaughn, D. L. (2017). *Applied multivariate statistical concepts.* Routledge.

Haig, B. D. (2018). *Method matters in psychology: Essays in applied philosophy of science.* Springer.

Hair, J. F., Black, W. C., Babin, B. J., & Anderson, R. E. (2019). *Multivariate data analysis* (8th ed.). Cengage Learning.

Haitovsky, Y. (1969). Multicollinearity in regression analysis: Comment. *The Review of Economics and Statistics, 51*(4), 486–489. https://doi.org/10.2307/1926450

Hancock, G. R., & Liu, M. (2012). Bootstrapping standard errors and data-model fit statistics in structural equation modeling. In R. H. Hoyle (Ed.), *Handbook of structural equation modeling* (pp. 296–306). Guilford.

Hancock, G. R., & Schoonen, R. (2015). Structural equation modeling: Possibilities for language learning researchers. *Language Learning, 65*(S1), 160–184. https://doi.org/10.1111/lang.12116

Harman, H. H. (1976). *Modern factor analysis* (3rd ed.). University of Chicago Press.

Hattori, M., Zhang, G., & Preacher, K. J. (2017). Multiple local solutions and geomin rotation. *Multivariate Behavioral Research, 52*(6), 720–731. https://doi.org/10.1080/00273171.2017.1361312

Hayashi, K., Bentler, P. M., & Yuan, K.-H. (2007). On the likelihood ratio test for the number of factors in exploratory factor analysis. *Structural Equation Modeling, 14*(3), 505–526. https://doi.org/10.1080/10705510701301891

Hayduk, L. A. (2014). Shame for disrespecting evidence: The personal consequences of insufficient respect for structural equation model testing. *BMC Medical Research Methodology, 14*(124), 1–24. https://doi.org/10.1186/1471-2288-14-124

Hayes, A. F., & Coutts, J. J. (2020). Use omega rather than Cronbach's alpha for estimating reliability. But… . *Communication Methods and Measures, 14*(1), 1–24. https://doi.org/10.1080/19312458.2020.1718629

Hayton, J. C., Allen, D. G., & Scarpello, V. (2004). Factor retention decisions in exploratory factor analysis: A tutorial on parallel analysis. *Organizational Research Methods, 7*(2), 191–205. https://doi.org/10.1177/1094428104263675

Heene, M., Hilbert, S., Draxler, C., Ziegler, M., & Bühner, M. (2011). Masking misfit in confirmatory factor analysis by increasing variances: A cautionary note on the usefulness of cutoff values of fit indices. *Psychological Methods, 16*(3), 319–336. https://doi.org/10.1037/a0024917

Hendrickson, A. E., & White, P. O. (1964). Promax: A quick method for rotation to oblique simple structure. *British Journal of Mathematical Psychology, 17*(1), 65–70. https://doi.org/10.1111/j.2044-8317.1964.tb00244.x

Henson, R. K., Hull, D. M., & Williams, C. S. (2010). Methodology in our education research culture: Toward a stronger collective quantitative proficiency. *Educational Researcher, 39*(3), 229–240. https://doi.org/10.3102/0013189X10365102

Henson, R. K., & Roberts, J. K. (2006). Use of exploratory factor analysis in published research: Common errors and some comment on improved practice. *Educational and Psychological Measurement, 66*(3), 393–416. https://doi.org/10.1177/0013164405282485

Hetzel, R. D. (1996). A primer on factor analysis with comments on patterns of practice and reporting. In B. Thompson (Ed.), *Advances in social science methodology* (Vol. 4, pp. 175–206). JAI Press.

Hoelzle, J. B., & Meyer, G. J. (2013). Exploratory factor analysis: Basics and beyond. In I. B. Weiner, J. A. Schinka, & W. F. Velicer (Eds.), *Handbook of psychology: Research methods in psychology* (Vol. 2, 2nd ed., pp. 164–188). Wiley.

Hogarty, K. Y., Hines, C. V., Kromrey, J. D., Ferron, J. M., & Mumford, K. R. (2005). The quality of factor solutions in exploratory factor analysis: The influence of sample size, communality, and overdetermination. *Educational and Psychological Measurement, 65*(2), 202–226. https://doi.org/10.1177/0013164404267287

Holgado-Tello, F. P., Chacón-Moscoso, S., Barbero-García, I., & Vila-Abad, E. (2010). Polychoric versus Pearson correlations in exploratory and confirmatory factor analysis of ordinal variables. *Quality & Quantity, 44*(1), 153–166. https://doi.org/10.1007/s11135-008-9190-y

Holzinger, K. J., & Harman, H. H. (1941). *Factor analysis; a synthesis of factorial methods.* University of Chicago Press.

Holzinger, K. J., & Swineford, F. (1937). The bi-factor method. *Psychometrika, 2*(1), 41–54. https://doi.org/10.1007/BF02287965

Holzinger, K. J., & Swineford, F. (1939). A study in factor analysis: The stability of a bifactor solution. *Supplementary Educational Monograph No. 48.* University of Chicago Press.

Hopwood, C. J., & Donnellan, M. B. (2010). How should the internal structure of personality inventories be evaluated? *Personality and Social Psychology Review, 14*(3), 332–346. https://doi.org/10.1177/1088868310361240

Horn, J. L. (1965). A rationale and test for the number of factors in factor analysis. *Psychometrika, 30*(2), 179–185. https://doi.org/10.1007/BF02289447

Howard, M. C. (2016). A review of exploratory factor analysis decisions and overview of current practices: What we are doing and how can we improve? *International Journal of Human–Computer Interaction, 32*(1), 51–62. https://doi.org/10.1080/10447318.2015.1087664

Hoyle, R. H. (2000). Confirmatory factor analysis. In H. E. A. Tinsley & S. D. Brown (Eds.), *Handbook of multivariate statistics and mathematical modeling* (pp. 465–497). Academic Press.

Hoyle, R. H., & Duvall, J. L. (2004). Determining the number of factors in exploratory and confirmatory factor analysis. In D. Kaplan (Ed.), *The Sage handbook of quantitative methodology for the social sciences* (pp. 301–315). Sage.

Hu, L., & Bentler, P. M. (1999). Cutoff criteria for fit indexes in covariance structure analysis: Conventional criteria versus new alternatives. *Structural Equation Modeling, 6*(1), 1–55. https://doi.org/10.1080/10705519909540118

Humphreys, L. G. (1982). The hierarchical factor model and general intelligence. In N. Hirschberg & L. G. Humphreys (Eds.), *Multivariate applications in the social sciences* (pp. 223–239). Erlbaum.

Hunsley, J., & Mash, E. J. (2007). Evidence-based assessment. *Annual Review of Clinical Psychology*, *3*(1), 29–51. https://doi.org/10.1146/annurev.clinpsy.3. 022806.091419

Hurley, A. E., Scandura, T. A., Schriesheim, C. A., Brannick, M. T., Seers, A., Vandenberg, R. J., & Williams, L. J. (1997). Exploratory and confirmatory factor analysis: Guidelines, issues, and alternatives. *Journal of Organizational Behavior*, *18*(6), 667–683. http://doi.org/cg5sf7

Hussey, I., & Hughes, S. (2020). Hidden invalidity among 15 commonly used measures in social and personality psychology. *Advances in Methods and Practices in Psychological Science*, *3*(2), 166–184. https://doi.org/10.1177/25152459198 82903

Hutchinson, S. R. (1998). The stability of post hoc model modifications in confirmatory factor analysis models. *Journal of Experimental Education*, *66*(4), 361–380. https://doi.org/10.1080/00220979809601406

Izquierdo, I., Olea, J., & Abad, F. J. (2014). Exploratory factor analysis in validation studies: Uses and recommendations. *Psicothema*, *26*(3), 395–400. https://doi.org/ 10.7334/psicothema2013.349

Jebb, A. T., Parrigon, S., & Woo, S. E. (2017). Exploratory data analysis as a foundation of inductive research. *Human Resource Management Review*, *27*(2), 265–276. https://doi.org/10.1016/j.hrmr.2016.08.003

Jennrich, R. I., & Bentler, P. M. (2011). Exploratory bi-factor analysis. *Psychometrika*, *76*(4), 537–549. https://doi.org/10.1007/s11336-011-9218-4

Jennrich, R. I., & Sampson, P. F. (1966). Rotation for simple loading. *Psychometrika*, *31*(3), 313–323. https://doi.org/10.1007/BF02289465

Johnson, R. L., & Morgan, G. B. (2016). *Survey scales: A guide to development, analysis, and reporting*. Guilford.

Kahn, J. H. (2006). Factor analysis in counseling psychology research, training, and practice: Principles, advances, and applications. *Counseling Psychologist*, *34*(5), 684– 718. https://doi.org/10.1177/0011000006286347

Kaiser, H. F. (1958). The varimax criterion for analytic rotation in factor analysis. *Psychometrika*, *23*(3), 187–200. https://doi.org/10.1007/BF02289233

Kaiser, H. F. (1974). An index of factorial simplicity. *Psychometrika*, *39*(1), 31–36. https://doi.org/10.1007/BF02291575

Kano, Y. (1997). Exploratory factor analysis with a common factor with two indicators. *Behaviormetrika*, *24*(2), 129–145. https://doi.org/10.2333/bhmk.24.129

Kanyongo, G. Y. (2005). Determining the correct number of components to extract from a principal components analysis: A Monte Carlo study of the accuracy of the scree plot. *Journal of Modern Applied Statistical Methods*, *4*(1), 120–133. https://doi. org/10.22237/jmasm/1114906380

Kline, P. (1991). *Intelligence: The psychometric view*. Routledge.

Kline, P. (1994). *An easy guide to factor analysis*. Routledge.

Kline, P. (2000). *A psychometrics primer*. Free Association Books.

Kline, R. B. (2012). Assumptions in structural equation modeling. In R. Hoyle (Ed.), *Handbook of structural equation modeling* (pp. 111–125). Guilford.

Kline, R. B. (2013). Exploratory and confirmatory factor analysis. In Y. Petscher, C. Schatschneider, & D. L. Compton (Eds.), *Applied quantitative analysis in education and the social sciences* (pp. 171–207). Routledge.

Koul, A., Becchio, C., & Cavallo, A. (2018). Cross-validation approaches for replicability in psychology. *Frontiers in Psychology*, *9*(1117), 1–4. https://doi.org/ 10.3389/fpsyg.2018.01117

Lai, K., & Green, S. B. (2016). The problem with having two watches: Assessment of fit when RMSEA and CFI disagree. *Multivariate Behavioral Research, 51*(2–3), 220–239. https://doi.org/10.1080/00273171.2015.1134306

Larsen, K. R., & Bong, C. H. (2016). A tool for addressing construct identity in literature reviews and meta-analyses. *MIS Quarterly, 40*(3), 529–551. https://doi.org/10.25300/MISQ/2016/40.3.01

Lawley, D. N., & Maxwell, A. E. (1963). *Factor analysis as a statistical method.* Butterworth.

Lawrence, F. R., & Hancock, G. R. (1999). Conditions affecting integrity of a factor solution under varying degrees of overextraction. *Educational and Psychological Measurement, 59*(4), 549–579. https://doi.org/10.1177/00131649921970026

Le, H., Schmidt, F. L., Harter, J. K., & Lauver, K. J. (2010). The problem of empirical redundancy of constructs in organizational research: An empirical investigation. *Organizational Behavior and Human Decision Processes, 112*(2), 112–125. https://doi.org/10.1016/j.obhdp.2010.02.003

Lee, C.-T., Zhang, G., & Edwards, M. C. (2012). Ordinary least squares estimation of parameters in exploratory factor analysis with ordinal data. *Multivariate Behavioral Research, 47*(2), 314–339. https://doi.org/10.1080/00273171.2012.658340

Lee, K., & Ashton, M. C. (2007). Factor analysis in personality research. In R. W. Robins, R. C. Fraley, & R. F. Krueger (Eds.), *Handbook of research methods in personality psychology* (pp. 424–443). Guilford.

Leech, N. L., & Goodwin, L. D. (2008). Building a methodological foundation: Doctoral-level methods courses in colleges of education. *Research in the Schools, 15*(1), 1–8.

Lei, P.-W., & Wu, Q. (2012). Estimation in structural equation modeling. In R. H. Hoyle (Ed.), *Handbook of structural equation modeling* (pp. 164–180). Guilford.

Lester, P. E., & Bishop, L. K. (2000). Factor analysis. In P. E. Lester & L. K. Bishop (Eds.), *Handbook of tests and measurement in education and the social sciences* (2nd ed., pp. 27–45). Scarecrow Press.

Leys, C., Klein, O., Dominicy, Y., & Ley, C. (2018). Detecting multivariate outliers: Use a robust variant of the Mahalanobis distance. *Journal of Experimental Social Psychology, 74*, 150–156. https://doi.org/10.1016/j.jesp.2017.09.011

Li, Y., Wen, Z., Hau, K.-T., Yuan, K.-H., & Peng, Y. (2020). Effects of cross-loadings on determining the number of factors to retain. *Structural Equation Modeling, 27*(6), 841–863. https://doi.org/10.1080/10705511.2020.1745075

Likert, R. (1932). A technique for the measurement of attitudes. *Archives of Psychology, 22*(140), 1–55.

Lim, S., & Jahng, S. (2019). Determining the number of factors using parallel analysis and its recent variants. *Psychological Methods, 24*(4), 452–467. https://doi.org/10.1037/met0000230

Little, T. D., Lindenberger, U., & Nesselroade, J. R. (1999). On selecting indicators for multivariate measurement and modeling with latent variables: When "good" indicators are bad and "bad" indicators are good. *Psychological Methods, 4*(2), 192–211. https://doi.org/10.1037/1082-989X.4.2.192

Little, T. D., Wang, E. W., & Gorrall, B. K. (2017). VIII. The past, present, and future of developmental methodology. *Monographs of the Society for Research in Child Development, 82*(2), 122–139. https://doi.org/10.1111/mono.12302

Liu, Y., Zumbo, B. D., & Wu, A. D. (2012). A demonstration of the impact of outliers on the decisions about the number of factors in exploratory factor analysis. *Educational and Psychological Measurement, 72*(2), 181–199. https://doi.org/10.1177/0013164411410878

Lloret, S., Ferreres, A., Hernández, A., & Tomás, I. (2017). The exploratory factor analysis of items: Guided analysis based on empirical data and software. *Anales de Psicologia, 33*(2), 417–432. https://doi.org/10.6018/analesps.33.2.270211

Longest, K. C. (2019). *Using stata for quantitative analysis* (3rd ed.). Sage.

Lorenzo-Seva, U., & Ferrando, P. J. (2019). A general approach for fitting pure exploratory bifactor models. *Multivariate Behavioral Research, 54*(1), 15–30. https://doi.org/10.1080/00273171.2018.1484339

Lorenzo-Seva, U., & Ferrando, P. J. (2021). Not positive definite correlation matrices in exploratory item factor analysis: Causes, consequences and a proposed solution. *Structural Equation Modeling, 28*(1), 138–147. https://doi.org/10.1080/10705511.2020.1735393

Lorenzo-Seva, U., Timmerman, M. E., & Kiers, H. A. L. (2011). The hull method for selecting the number of common factors. *Multivariate Behavioral Research, 46*(2), 340–364. https://doi.org/10.1080/00273171.2011.564527

Lozano, L. M., Garcia-Cueto, E., & Muniz, J. (2008). Effect of the number of response categories on the reliability and validity of rating scales. *Methodology, 4*(2), 73–79. https://doi.org/10.1027/1614-2241.4.2.73

Lubinski, D., & Dawis, R. V. (1992). Aptitudes, skills, and proficiencies. In M. D. Dunnette & L. M. Hough (Eds.), *Handbook of industrial and organizational psychology* (2nd ed., Vol. 3, pp. 1–59). Consulting Psychology Press.

MacCallum, R. C. (2003). Working with imperfect models. *Multivariate Behavioral Research, 38*(1), 113–139. https://doi.org/10.1207/S15327906MBR3801_5

MacCallum, R. C. (2009). Factor analysis. In R. E. Millsap & A. Maydeu-Olivares (Eds.), *Sage handbook of quantitative methods in psychology* (pp. 123–147). Sage.

MacCallum, R. C., & Austin, J. T. (2000). Applications of structural equation modeling in psychological research. *Annual Review of Psychology, 51*(1), 201–226. https://doi.org/10.1146/annurev.psych.51.1.201

MacCallum, R. C., Browne, M. W., & Cai, L. (2007). Factor analysis models as approximations. In R. Cudeck & R. C. MacCallum (Eds.), *Factor analysis at 100: Historical developments and future directions* (pp. 153–175). Erlbaum.

MacCallum, R. C., Roznowski, M., & Necowitz, L. B. (1992). Model modifications in covariance structure analysis: The problem of capitalization on chance. *Psychological Bulletin, 111*(3), 490–504. https://doi.org/10.1037/0033-2909.111.3.490

MacCallum, R. C., Widaman, K. F., Preacher, K. J., & Hong, S. (2001). Sample size in factor analysis: The role of model error. *Multivariate Behavioral Research, 36*(4), 611–637. https://doi.org/10.1207/S15327906MBR3604_06

MacCallum, R. C., Widaman, K. F., Zhang, S., & Hong, S. (1999). Sample size in factor analysis. *Psychological Methods, 4*(1), 84–99. https://doi.org/10.1037/1082-989X.4.1.84

Malone, P. S., & Lubansky, J. B. (2012). Preparing data for structural equation modeling: Doing your homework. In R. H. Hoyle (Ed.), *Handbook of structural equation modeling* (pp. 263–276). Guilford.

Mansolf, M., & Reise, S. P. (2015). Local minima in exploratory bifactor analysis. *Multivariate Behavioral Research, 50*(6), 738. https://doi.org/10.1080/00273171.2015.1121127

Mansolf, M., & Reise, S. P. (2016). Exploratory bifactor analysis: The Schmid-Leiman orthogonalization and Jennrich-Bentler analytic rotations. *Multivariate Behavioral Research, 51*(5), 698–717. https://doi.org/10.1080/00273171.2016.1215898

Mardia, K. V. (1970). Measures of multivariate skewness and kurtosis with applications. *Biometrika, 57*(3), 519–530. https://doi.org/10.1093/biomet/57.3.519

Marsh, H. W. (1990). *Self-Description Questionnaire–II manual.* University of Western Sydney, Macarthur.

Marsh, H. W., Muthen, B., Asparouhov, T., Ludtke, O., Robitzsch, A., Morin, A. J. S., & Trautwein, U. (2009). Exploratory structural equation modeling, integrating CFA and EFA: Application to students' evaluations of university teaching. *Structural Equation Modeling, 16*(3), 439–476. https://doi.org/10.1080/10705510903008220

Matsunaga, M. (2010). How to factor-analyze your data right: Do's, don'ts, and how-to's. *International Journal of Psychological Research, 3*(1), 97–110. https://doi.org/10.21500/20112084.854

Maydeu-Olivares, A. (2017). Assessing the size of model misfit in structural equation models. *Psychometrika, 82*(3), 533–558. https://doi.org/10.1007/s11336-016-9552-7

McArdle, J. J. (2011). Some ethical issues in factor analysis. In A. T. Panter & S. K. Sterba (Eds.), *Handbook of ethics in quantitative methodology* (pp. 313–339). Routledge.

McClain, A. L. (1996). Hierarchical analytic methods that yield different perspectives on dynamics: Aids to interpretation. In B. Thompson (Ed.), *Advances in social science methodology* (Vol. 4, pp. 229–240). JAI Press.

McCoach, D. B., Gable, R. K., & Madura, J. P. (2013). *Instrument development in the affective domain: School and corporate applications.* Springer.

McCroskey, J. C., & Young, T. J. (1979). The use and abuse of factor analysis in communication research. *Human Communication Research, 5*(4), 375–382. https://doi.org/10.1111/j.1468-2958.1979.tb00651.x

McDonald, R. P. (1985). *Factor analysis and related methods.* Erlbaum.

McDonald, R. P. (1999). *Test theory: A unified approach.* Erlbaum.

McDonald, R. P. (2010). Structural models and the art of approximation. *Perspectives on Psychological Science, 5*(6), 675–686. https://doi.org/10.1177/1745691610388766

Meehl, P. E. (1990). Why summaries of research on psychological theories are often uninterpretable. *Psychological Reports, 66*(1), 195–244. https://doi.org/10.2466/pr0.1990.66.1.195

Meehl, P. E. (2006). The power of quantitative thinking. In N. G. Waller, L. J. Yonce, W. M. Grove, D. Faust, & M. F. Lenzenweger (Eds.), *A Paul Meehl reader: Essays on the practice of scientific psychology* (pp. 433–444). Erlbaum.

Mehmetoglu, M., & Jakobsen, T. G. (2017). *Applied statistics using Stata: A guide for the social sciences.* Sage.

Mertler, C. A., & Vannatta, R. A. (2001). *Advanced and multivariate statistical methods: Practical application and interpretation.* Pyrczak Publishing.

Messick, S. (1995). Validity of psychological assessment. *American Psychologist, 50*(9), 741–749. https://doi.org/10.1037/0003-066X.50.9.741

Mitchell, M. N. (2020). *Data management using Stata: A practical handbook* (end ed.). Stata Press.

Molina, J., Servera, M., & Burns, G. L. (2020). Structure of ADHD/ODD symptoms in Spanish preschool children: Dangers of confirmatory factor analysis for evaluation of rating scales. *Assessment, 27*(8), 1748–1757. https://doi.org/10.1177/1073191119839140

Montoya, A. K., & Edwards, M. C. (2021). The poor fit of model fit for selecting number of factors in exploratory factor analysis for scale evaluation. *Educational and Psychological Measurement, 81*(3), 413-440. https://doi.org/10.1177/0013164420942899

Morin, A. J. S., Arens, A. K., Tran, A., & Caci, H. (2016). Exploring sources of construct-relevant multidimensionality in psychiatric measurement: A tutorial and illustration using the composite scale of morningness. *International Journal of Methods in Psychiatric Research, 25*(4), 277–288. https://doi.org/10.1002/mpr.1485

Morin, A. J. S., Myers, N. D., & Lee, S. (2020). Modern factor analytic techniques. In G. Tenenbaum & R. C. Eklund (Eds.), *Handbook of sport psychology* (4th ed., pp. 1044–1073). Wiley.

Morrison, J. T. (2009). Evaluating factor analysis decisions for scale design in communication research. *Communication Methods and Measures, 3*(4), 195–215. https://doi.org/10.1080/19312450903378917

Mucherah, W., & Finch, H. (2010). The construct validity of the Self Description Questionnaire on high school students in Kenya. *International Journal of Testing, 10*(2), 166–184. https://doi.org/10.1080/15305051003739904

Mueller, R. O., & Hancock, G. R. (2019). Structural equation modeling. In G. R. Hancock, L. M. Stapleton, & R. O. Mueller (Eds.), *The reviewer's guide to quantitative methods in the social sciences* (2nd ed., pp. 445–456). Routledge.

Mulaik, S. A. (1987). A brief history of the philosophical foundations of exploratory factor analysis. *Multivariate Behavioral Research, 22*(3), 267–305. https://doi.org/10.1207/s15327906mbr2203_3

Mulaik, S. A. (2010). *Foundations of factor analysis* (2nd ed.). Chapman & Hall/CRC.

Mulaik, S. A. (2018). Fundamentals of common factor analysis. In R. Irwing, T. Booth, & D. J. Hughes (Eds.), *The Wiley handbook of psychometric testing: A multidisciplinary reference on survey, scale and test development* (pp. 211–251). Wiley.

Mundfrom, D. J., Shaw, D. G., & Ke, T. L. (2005). Minimum sample size recommendations for conducting factor analyses. *International Journal of Testing, 5*(2), 159–168. https://doi.org/10.1207/s15327574ijt0502_4

Murphy, K. R., & Aguinis, H. (2019). HARKing: How badly can cherry-picking and question trolling produce bias in published results? *Journal of Business and Psychology, 34*(1), 1–17. https://doi.org/10.1007/s10869-017-9524-7

Mvududu, N. H., & Sink, C. A. (2013). Factor analysis in counseling research and practice. *Counseling Outcome Research and Evaluation, 4*(2), 75–98. https://doi.org/10.1177/2150137813494766

Nasser, F., Benson, J., & Wisenbaker, J. (2002). The performance of regression-based variations of the visual scree for determining the number of common factors. *Educational and Psychological Measurement, 62*(3), 397–419. https://doi.org/10.1177/0016440206200300

Nesselroade, J. R. (1994). Exploratory factor analysis with latent variables and the study of processes of development and change. In A. von Eye & C. C. Clogg (Eds.), *Latent variables analysis: Applications for developmental research* (pp. 131–154). Sage.

Newman, D. A. (2014). Missing data: Five practical guidelines. *Organizational Research Methods, 17*(4), 372–411. https://doi.org/10.1177/1094428114548590

Nichols, J. Q. V. A., Shoulberg, E. K., Garner, A. A., Hoza, B., Burt, K. B., Murray-Close, D., & Arnold, L. E. (2017). Exploration of the factor structure of ADHD in adolescence through self, parent, and teacher reports of symptomatology. *Journal of Abnormal Child Psychology, 45*(3), 625–641. https://doi.org/10.1007/s10802-016-0183-3

Norman, G. R., & Streiner, D. L. (2014). *Biostatistics: The bare essentials* (4th ed.). People's Medical Publishing.

Norris, M., & Lecavalier, L. (2010). Evaluating the use of exploratory factor analysis in developmental disability psychological research. *Journal of Autism and Developmental Disorders, 40*(1), 8–20. https://doi.org/10.1007/s10803-009-0816-2

Nunnally, J. C., & Bernstein, I. H. (1994). *Psychometric theory* (3rd ed.). McGraw-Hill.

Onwuegbuzie, A. J., & Daniel, L. G. (2002). Uses and misuses of the correlation coefficient. *Research in the Schools, 9*(1), 73–90.

Open Science Collaboration. (2015). Estimating the reproducibility of psychological science. *Science, 349*(6251), 1–8. https://doi.org/10.1126/science.aac4716

Orcan, F. (2018). Exploratory and confirmatory factor analysis: Which one to use first? *Journal of Measurement and Evaluation in Education and Psychology, 9*(4), 414–421. https://doi.org/10.21031/epod.394323

Osborne, J. W. (2014). *Best practices in exploratory factor analysis.* CreateSpace Independent Publishing.

Osborne, J. W., & Banjanovic, E. S. (2016). *Exploratory factor analysis with SAS.* SAS Institute.

Osborne, J. W., Costello, A. B., & Kellow, J. T. (2007). Best practices in exploratory factor analysis. In J. W. Osborne (Ed.), *Best practices in quantitative methods* (pp. 86–99). Sage.

Osborne, J. W., & Fitzpatrick, D. C. (2012). Replication analysis in exploratory factor analysis: What it is and why it makes your analysis better. *Practical Assessment, Research & Evaluation, 17*(15), 1–8. https://doi.org/10.7275/h0bd-4d11

Panter, A. T., Swygert, K. A., Dahlstrom, W. G., & Tanaka, J. S. (1997). Factor analytic approaches to personality item-level data. *Journal of Personality Assessment, 68*(3), 561–589. https://doi.org/10.1207/s15327752jpa6803_6

Park, H. S., Dailey, R., & Lemus, D. (2002). The use of exploratory factor analysis and principal components analysis in communication research. *Human Communication Research, 28*(4), 562–577. https://doi.org/10.1111/j.1468-2958.2002.tb00824.x

Pearson, R. H., & Mundfrom, D. J. (2010). Recommended sample size for conducting exploratory factor analysis on dichotomous data. *Journal of Modern Applied Statistical Methods, 9*(2), 359–368. https://doi.org/10.22237/jmasm/1288584240

Peres-Neto, P., Jackson, D., & Somers, K. (2005). How many principal components? Stopping rules for determining the number of non-trivial axes revisited. *Computational Statistics Data Analysis, 49*(4), 974–997. https://doi.org/10.1016/j.csda.2004.06.015

Peterson, C. (2017). Exploratory factor analysis and theory generation in psychology. *Review of Philosophy and Psychology, 8*(3), 519–540. https://doi.org/10.1007/s13164-016-0325-0

Peterson, R. A. (2000). A meta-analysis of variance accounted for and factor loadings in exploratory factor analysis. *Marketing Letters, 11*(3), 261–275. https://doi.org/10.1023/A:1008191211004

Pett, M. A., Lackey, N. R., & Sullivan, J. J. (2003). *Making sense of factor analysis.* Sage.

Pituch, K. A., & Stevens, J. P. (2016). *Applied multivariate statistics for the social sciences* (6th ed.). Routledge.

Platt, J. R. (1964). Strong inference. *Science, 146*(3642), 347–353. https://doi.org/10.1126/science.146.3642.347

Plonsky, L., & Gonulal, T. (2015). Methodological synthesis in quantitative L2 research: A review of reviews and a case study of exploratory factor analysis. *Language Learning, 65*(S1), 9–36. https://doi.org/10.1111/lang.12111

Podsakoff, P. M., MacKenzie, S. B., & Podsakoff, N. P. (2012). Sources of method bias in social science research and recommendations on how to control it. *Annual Review of Psychology, 63*(1), 539–569. https://doi.org/10.1146/annurev-psych-120710-100452

Popper, K. (2002). *Conjectures and refutations: The growth of scientific knowledge.* Routledge.

Preacher, K. J., & MacCallum, R. C. (2003). Repairing Tom Swift's electric factor analysis machine. *Understanding Statistics, 2*(1), 13–43. https://doi.org/10.1207/S15328031US0201_02

Preacher, K. J., & Merkle, E. C. (2012). The problem of model selection uncertainty in structural equation modeling. *Psychological Methods, 17*(1), 1–14. https://doi. org/10.1037/a0026804

Preacher, K. J., Zhang, G., Kim, C., & Mels, G. (2013). Choosing the optimal number of factors in exploratory factor analysis: A model selection perspective. *Multivariate Behavioral Research, 48*(1), 28–56. https://doi.org/10.1080/ 00273171.2012.710386

Puth, M.-T., Neuhäuser, M., & Ruxton, G. D. (2015). Effective use of Spearman's and Kendall's correlation coefficients for association between two measured traits. *Animal Behaviour, 102*, 77–84. https://doi.org/10.1016/j.anbehav.2015.01.010

R Core Team. (2020). *R: A language and environment for statistical computing.* R Foundation for Statistical Computing.

Raîche, G., Walls, T. A., Magis, D., Riopel, M., & Blais, J.-G. (2013). Non-graphical solutions for Cattell's scree test. *Methodology, 9*(1), 23–29. https://doi.org/ 10.1027/1614-2241/a000051

Ramlall, I. (2017). *Applied structural equation modelling for researchers and practitioners.* Emerald Group Publishing.

Reio, T. G., & Shuck, B. (2015). Exploratory factor analysis: Implications for theory, research, and practice. *Advances in Developing Human Resources, 17*(1), 12–25. https://doi.org/10.1177/1523422314559804

Reise, S. P. (2012). The rediscovery of bifactor measurement models. *Multivariate Behavioral Research, 47*(5), 667–696. https://doi.org/10.1080/00273171.2012.715555

Reise, S. P., Bonifay, W., & Haviland, M. G. (2018). Bifactor modelling and the evaluation of scale scores. In P. Irwing, T. Booth, & D. J. Hughes (Eds.), *The Wiley handbook of psychometric testing: A multidisciplinary reference on survey, scale and test development* (pp. 677–707). Wiley.

Reise, S. P., Moore, T. M., & Haviland, M. G. (2010). Bifactor models and rotations: Exploring the extent to which multidimensional data yield univocal scale scores. *Journal of Personality Assessment, 92*(6), 544–559. https://doi.org/ 10.1080/00223891.2010.496477

Reise, S. P., Waller, N. G., & Comrey, A. L. (2000). Factor analysis and scale revision. *Psychological Assessment, 12*(3), 287–297. https://doi.org/10.1037/ 1040-3590.12.3.287

Rencher, A. C., & Christensen, W. F. (2012). *Methods of multivariate analysis* (3rd ed.). Wiley.

Revelle, W. (2016). An introduction to psychometric theory with applications in R. Retrieved from http://personality-project.org/r/book/

Revelle, W. and Rocklin, T. (1979). Very simple structure: Alternative procedure for estimating the optimal number of interpretable factors. *Multivariate Behavioral Research, 14*(4), 403–414. https://doi.org/10.1207/s15327906mbr1404_2

Rhemtulla, M., Brosseau-Liard, P. E., & Savalei, V. (2012). When can categorical variables be treated as continuous? A comparison of continuous and categorical SEM estimation methods under suboptimal conditions. *Psychological Methods, 17*(3), 354–373. https://doi.org/10.1037/a0029315

Rhemtulla, M., van Bork, R., & Borsboom, D. (2020). Worse than measurement error: Consequences of inappropriate latent variable measurement models. *Psychological Methods, 25*(1), 30–45. https://doi.org/10.1037/met0000220

Rigdon, E. E., Becker, J.-M., & Sarstedt, M. (2019). Factor indeterminacy as metrological uncertainty: Implications for advancing psychological measurement. *Multivariate Behavioral Research, 54*(3), 429–443. https://doi.org/10.1080/ 00273171.2018.1535420

Roberson, R. B., Elliott, T. R., Chang, J. E., & Hill, J. N. (2014). Exploratory factor analysis in rehabilitation psychology: A content analysis. *Rehabilitation Psychology, 59*(4), 429–438. https://doi.org/10.1037/a0037899

Roberts, S., & Pashler, H. (2000). How persuasive is a good fit? A comment on theory testing. *Psychological Review, 107*(2), 358–367. https://doi.org/10.1037/0033-295X.107.2.358

Rockwell, R. C. (1975). Assessment of multicollinearity: The Haitovsky test of the determinant. *Sociological Methods & Research, 3*(3), 308–320. https://doi.org/10.1177/004912417500300304

Rodriguez, A., Reise, S. P., & Haviland, M. G. (2016). Applying bifactor statistical indices in the evaluation of psychological measures. *Journal of Personality Assessment, 98*(3), 223–237. https://doi.org/10.1080/00223891.2015.1089249

Ropovik, I. (2015). A cautionary note on testing latent variable models. *Frontiers in Psychology, 6*(1715), 1–8. https://doi.org/10.3389/fpsyg.2015.01715

Roth, P. L. (1994). Missing data: A conceptual review for applied psychologists. *Personnel Psychology, 47*(3), 537–560. https://doi.org/10.1111/j.1744-6570.1994.tb01736.x

Rouquette, A., & Falissard, B. (2011). Sample size requirements for the validation of psychiatric scales. *International Journal of Methods in Psychiatric Research, 20*(4), 235–249. https://doi.org/10.1002/mpr.352

Rubin, D. B. (1976). Inference and missing data. *Biometrika, 63*(3), 581–592. https://doi.org/10.1093/biomet/63.3.581

Rubin, M. (2017). When does HARKing hurt? Identifying when different types of undisclosed post hoc hypothesizing harm scientific progress. *Review of General Psychology, 21*(4), 308–320. https://doi.org/10.1037/gpr0000128

Rummel, R. J. (1967). Understanding factor analysis. *Journal of Conflict Resolution, 11*(4), 444–480. https://doi.org/10.1177/002200276701100405

Rummel, R. J. (1970). *Applied factor analysis*. Northwestern University Press.

Ruscio, J., & Roche, B. (2012). Determining the number of factors to retain in an exploratory factor analysis using comparison data of known factorial structure. *Psychological Assessment, 24*(2), 282–292. https://doi.org/10.1037/a0025697

Russell, D. W. (2002). In search of underlying dimensions: The use (and abuse) of factor analysis in Personality and Social Psychology Bulletin. *Personality and Social Psychology Bulletin, 28*(12), 1629–1646. https://doi.org/10.1177/014616702237645

Sakaluk, J. K., & Short, S. D. (2017). A methodological review of exploratory factor analysis in sexuality research: Used practices, best practices, and data analysis resources. *Journal of Sex Research, 54*(1), 1–9. https://doi.org/10.1080/00224499.2015.1137538

Saris, W. E., Satorra, A., & van der Veld, W. M. (2009). Testing structural equation models or detection of misspecifications? *Structural Equation Modeling, 16*(4), 561–582. https://doi.org/10.1080/10705510903203433

Sass, D. A. (2010). Factor loading estimation error and stability using exploratory factor analysis. *Educational and Psychological Measurement, 70*(4), 557–577. https://doi.org/10.1177/0013164409355695

Sass, D. A., & Schmitt, T. A. (2010). A comparative investigation of rotation criteria within exploratory factor analysis. *Multivariate Behavioral Research, 45*(1), 73–103. https://doi.org/10.1080/00273170903504810

Savalei, V. (2012). The relationship between root mean square error of approximation and model misspecification in confirmatory factor analysis models. *Educational and Psychological Measurement, 72*(6), 910–932. https://doi.org/10.1177/0013164412452564

Schmid, J., & Leiman, J. M. (1957). The development of hierarchical factor solutions. *Psychometrika, 22*(1), 53–61. https://doi.org/10.1007/BF02289209

Schmitt, T. A. (2011). Current methodological considerations in exploratory and confirmatory factor analysis. *Journal of Psychoeducational Assessment, 29*(4), 304–321. https://doi.org/10.1177/0734282911406653

Schmitt, T. A., & Sass, D. A. (2011). Rotation criteria and hypothesis testing for exploratory factor analysis: Implications for factor pattern loadings and interfactor correlations. *Educational and Psychological Measurement, 71*(1), 95–113. https://doi.org/10.1177/0013164410387348

Schmitt, T. A., Sass, D. A., Chappelle, W., & Thompson, W. (2018). Selecting the "best" factor structure and moving measurement validation forward: An illustration. *Journal of Personality Assessment, 100*(4), 345–362. https://doi.org/10.1080/00223891.2018.1449116

Schönbrodt, F. D., & Perugini, M. (2013). At what sample size do correlations stabilize? *Journal of Research in Personality, 47*(5), 609–612. https://doi.org/10.1016/j.jrp.2013.05.009

Schumacker, R. E., & Lomax, R. G. (2004). *A beginner's guide to structural equation modeling* (2nd ed.). Erlbaum.

Schwarz, G. (1978). Estimating the dimension of a model. *Annals of Statistics, 6*(2), 461–464. https://doi.org/10.1214/aos/1176344136

Sellbom, M., & Tellegen, A. (2019). Factor analysis in psychological assessment research: Common pitfalls and recommendations. *Psychological Assessment, 31*(12), 1428–1441. https://doi.org/10.1037/pas0000623

Shaffer, J. A., DeGeest, D., & Li, A. (2016). Tackling the problem of construct proliferation: A guide to assessing the discriminant validity of conceptually related constructs. *Organizational Research Methods, 19*(1), 80–110. https://doi.org/10.1177/1094428115598239

Sheskin, D. J. (2011). *Handbook of parametric and nonparametric statistical procedures* (5th ed.). CRC Press.

Shi, D., & Olivares, A. (2020). The effect of estimation methods on SEM fit indices. *Educational and Psychological Measurement, 80*(3), 421–445. https://doi.org/10.1177/0013164419885164

Simmons, J. P., Nelson, L. D., & Simonsohn, U. (2011). False-positive psychology: Undisclosed flexibility in data collection and analysis allows presenting anything as significant. *Psychological Science, 22*(11), 1359–1366. https://doi.org/10.1177/0956797611417632

Simms, L. J., & Watson, D. (2007). The construct validation approach to personality scale construction. In R. W. Roberts, R. C. Fraley, & R. F. Krueger (Eds.), *Handbook of research methods in personality psychology* (pp. 240–258). Guilford.

Spearman, C. (1904). "General intelligence," objectively determined and measured. *American Journal of Psychology, 15*(2), 201–293. https://doi.org/10.1037/11491-006

Spector, P. E., Van Katwyk, P. T., Brannick, M. T., & Chen, P. Y. (1997). When two factors don't reflect two constructs: How item characteristics can produce artifactual factors. *Journal of Management, 23*(5), 659–677. https://doi.org/10.1177/014920639702300503

Spurgeon, S. L. (2017). Evaluating the unintended consequences of assessment practices: Construct irrelevance and construct underrepresentation. *Measurement and Evaluation in Counseling and Development, 50*(4), 275–281. https://doi.org/10.1080/07481756.2017.1339563

StataCorp. (2019). *Stata statistical software: Release 16*. StataCorp LLC.

Steiger, J. H. (1990). Structural model evaluation and modification: An interval esti-mation approach. *Multivariate Behavioral Research*, *25*(2), 173–180. https://doi.org/10.1207/s15327906mbr2502_4

Steiger, J. H. (2001). Driving fast in reverse: The relationship between software development, theory, and education in structural equation modeling. *Journal of the American Statistical Association*, *96*(453), 331–338. https://doi.org/10.1198/016214501750332893

Stevens, S. S. (1946). On the theory of scales of measurement. *Science*, *103*(2684), 677–680. https://doi.org/10.1126/science.103.2684.677

Stewart, D. W. (1981). The application and misapplication of factor analysis in marketing research. *Journal of Marketing Research*, *18*(1), 51–62. https://doi.org/10.1177/002224378101800105

Stewart, D. W. (2001). Factor analysis. *Journal of Consumer Psychology*, *10*(1–2), 75–82. https://onlinelibrary.wiley.com/doi/10.1207/S15327663JCP1001%262_07

Streiner, D. L. (1994). Figuring out factors: The use and misuse of factor ana-lysis. *Canadian Journal of Psychiatry*, *39*(3), 135–140. https://doi.org/10.1177/070674379403900303

Streiner, D. L. (1998). Factors affecting reliability of interpretations of scree plots. *Psychological Reports*, *83*(2), 687–694. https://doi.org/10.2466/pr0.1998.83.2.687

Streiner, D. L. (2018). Commentary no. 26: Dealing with outliers. *Journal of Clinical Psychopharmacology,38*(3),170–171.https://doi.org/10.1097/jcp.0000000000000865

Tabachnick, B. G., & Fidell, L. S. (2019). *Using multivariate statistics* (7th ed.). Pearson.

Tarka, P. (2018). An overview of structural equation modeling: Its beginnings, his-torical development, usefulness and controversies in the social sciences. *Quality & Quantity*, *52*(1), 313–354. https://doi.org/10.1007/s11135-017-0469-8

Tataryn, D. J., Wood, J. M., & Gorsuch, R. L. (1999). Setting the value of k in promax: A Monte Carlo study. *Educational and Psychological Measurement*, *59*(3), 384–391. https://doi.org/10.1177/00131649921969938

Themessl-Huber, M. (2014). Evaluation of the X^2-statistic and different fit-indices under misspecified number of factors in confirmatory factor analysis. *Psychological Test and Assessment Modeling*, *56*(3), 219–236.

Thompson, B. (2004). *Exploratory and confirmatory factor analysis: Understanding concepts and applications*. American Psychological Association.

Thomson, G. (1950). *The factorial analysis of human ability* (4th ed.). University of London Press.

Thurstone, L. L. (1931). Multiple factor analysis. *Psychological Review*, *38*(5), 406–427. https://doi.org/10.1037/h0069792

Thurstone, L. L. (1935). *The vectors of mind: Multiple-factor analysis for the isolation of primary traits*. University of Chicago Press.

Thurstone, L. L. (1937). Current misuse of the factorial methods. *Psychometrika*, *2*(2), 73–76. https://doi.org/10.1007/BF02288060

Thurstone, L. L. (1940). Current issues in factor analysis. *Psychological Bulletin*, *37*(4), 189–236. https://doi.org/10.1037/h0059402

Thurstone, L. L. (1947). *Multiple factor analysis*. University of Chicago Press.

Tinsley, H. E. A., & Tinsley, D. J. (1987). Uses of factor analysis in counseling psych-ology research. *Journal of Counseling Psychology*, *34*(4), 414–424. https://doi.org/10.1037/0022-0167.34.4.414

Tomarken, A. J., & Waller, N. G. (2003). Potential problems with "well fitting" models. *Journal of Abnormal Psychology*, *112*(4), 578–598. https://doi.org/10.1037/0021-843X.112.4.578

Tomarken, A. J., & Waller, N. G. (2005). Structural equation modeling: Strengths, limitations, and misconceptions. *Annual Review of Clinical Psychology*, *1*(1), 31–65. https://doi.org/10.1146/annurev.clinpsy.1.102803.144239

Tukey, J. W. (1980). We need both exploratory and confirmatory. *American Statistician*, *34*(1), 23–25. https://doi.org/10.1080/00031305.1980.10482706

van der Eijk, C., & Rose, J. (2015). Risky business: Factor analysis of survey data – Assessing the probability of incorrect dimensionalisation. *PLoS One*, *10*(3), e0118900. https://doi.org/10.1371/journal.pone.0118900

van Driel, O. P. (1978). On various causes of improper solutions in maximum likelihood factor analysis. *Psychometrika*, *43*(2), 225–243. https://doi.org/10.1007/BF02293865

Velicer, W. F. (1976). Determining the number of components from the matrix of partial correlations. *Psychometrika*, *41*(3), 321–327. https://doi.org/10.1007/BF02293557

Velicer, W. F., Eaton, C. A., & Fava, J. L. (2000). Construct explication through factor or component analysis: A review and evaluation of alternative procedures for determining the number of factors or components. In R. D. Goffin & E. Helmes (Eds.), *Problems and solutions in human assessment: Honoring Douglas N. Jackson at seventy* (pp. 41–71). Kluwer Academic Publishers.

Velicer, W. F., & Fava, J. L. (1998). Effects of variable and subject sampling on factor pattern recovery. *Psychological Methods*, *3*(2), 231–251. https://doi.org/10.1037/1082-989X.3.2.231

Velicer, W. F., & Jackson, D. N. (1990). Component analysis versus common factor analysis: Some issues in selecting an appropriate procedure. *Multivariate Behavioral Research*, *25*(1), 1–28. https://doi.org/10.1207/s15327906mbr2501_1

Vernon, P. E. (1961). *The structure of human abilities* (2nd ed.). Methuen.

Wainer, H. (1976). Estimating coefficients in linear models: It don't make no nevermind. *Psychological Bulletin*, *83*(2), 213–217. https://doi.org/10.1037/0033-2909.83.2.213

Walkey, F., & Welch, G. (2010). *Demystifying factor analysis: How it works and how to use it*. Xlibris.

Walsh, B. D. (1996). A note on factors that attenuate the correlation coefficient and its analogs. In B. Thompson (Ed.), *Advances in social science methodology* (Vol. 4, pp. 21–31). JAI Press.

Wang, L. L., Watts, A. S., Anderson, R. A., & Little, T. D. (2013). Common fallacies in quantitative research methodology. In T. D. Little (Ed.), *Oxford handbook of quantitative methods: Statistical analysis* (Vol. 2, pp. 718–758). Oxford University Press.

Warne, R. T., & Burningham, C. (2019). Spearman's *g* found in 31 non-Western nations: Strong evidence that *g* is a universal phenomenon. *Psychological Bulletin*, *145*(3), 237–272. http://dx.doi.org/10.1037/bul0000184

Warner, R. M. (2007). *Applied statistics: From bivariate through multivariate techniques*. Sage.

Wasserstein, R. L., & Lazar, N. A. (2016). The ASA's statement on *p*-values: Context, process, and purpose. *The American Statistician*, *70*(2), 129–133. https://doi.org/10.1080/00031305.2016.1154108

Watkins, M. W. (2006). Orthogonal higher-order structure of the Wechsler Intelligence Scale for Children–Fourth Edition. *Psychological Assessment, 18*(1), 123–125. https://doi.org/10.1037/1040-3590.18.1.123

Watkins, M. W. (2009). Errors in diagnostic decision making and clinical judgment. In T. B. Gutkin & C. R. Reynolds (Eds.), *Handbook of school psychology* (4th ed., pp. 210–229). Wiley.

Watkins, M. W. (2017). The reliability of multidimensional neuropsychological measures: From alpha to omega. *The Clinical Neuropsychologist, 31*(6–7), 1113–1126. https://doi.org/10.1080/13854046.2017.1317364

Watkins, M. W. (2018). Exploratory factor analysis: A guide to best practice. *Journal of Black Psychology, 44*(3), 219–246. https://doi.org/10.1177/0095798418771807

Watkins, M. W., & Browning, L. J. (2015). The Baylor revision of the Motivation to Read Survey (B-MRS). *Research and Practice in the Schools, 3*(1), 37–50.

Watkins, M. W., & Canivez, G. L. (2021). Assessing the psychometric utility of IQ scores: A tutorial using the Wechsler Intelligence Scale for Children–Fifth Edition. *School Psychology Review.* https://doi.org/10.1080/2372966X.2020.1816804

Watkins, M. W., Greenawalt, C. G., & Marcell, C. M. (2002). Factor structure of the Wechsler Intelligence Scale for Children: Third edition among gifted students. *Educational and Psychological Measurement, 62*(1), 164–172. https://doi.org/10.1177/0013164402062001011

Watson, J. C. (2017). Establishing evidence for internal structure using exploratory factor analysis. *Measurement and Evaluation in Counseling and Development, 50*(4), 232–238. https://doi.org/10.1080/07481756.2017.1336931

Weaver, B., & Maxwell, H. (2014). Exploratory factor analysis and reliability analysis with missing data: A simple method for SPSS users. *Quantitative Methods for Psychology, 10*(2), 143–152. https://doi.org/10.20982/tqmp.10.2.p143

Wegener, D. T., & Fabrigar, L. R. (2000). Analysis and design for nonexperimental data. In H. T. Reis & C. M. Judd (Eds.), *Handbook of research methods in social and personality psychology* (pp. 412–450). Cambridge University Press.

Wetzel, E., & Roberts, B. W. (2020). Commentary on Hussey and Hughes (2020): Hidden invalidity among 15 commonly used measures in social and personality psychology. *Advances in Methods and Practices in Psychological Science, 3*(4), 505–508. https://doi.org/10.1177/2515245920957618

Widaman, K. F. (1993). Common factor analysis versus principal component analysis: Differential bias in representing model parameters? *Multivariate Behavioral Research, 28*(3), 263–311. https://doi.org/10.1207/s15327906mbr2803_1

Widaman, K. F. (2012). Exploratory factor analysis and confirmatory factor analysis. In H. Cooper (Ed.), *APA handbook of research methods in psychology: Data analysis and research publication* (Vol. 3, pp. 361–389). American Psychological Association.

Widaman, K. F. (2018). On common factor and principal component representations of data: Implications for theory and for confirmatory replications. *Structural Equation Modeling, 25*(6), 829–847. https://doi.org/10.1080/10705511.2018.1478730

Williams, B., Onsman, A., & Brown, T. (2010). Exploratory factor analysis: A five-step guide for novices. *Journal of Emergency Primary Health Care, 8*(3), 1–13. https://doi.org/10.33151/ajp.8.3.93

Wolf, E. J., Harrington, K. M., Clark, S. L., & Miller, M. W. (2013). Sample size requirements for structural equation models: An evaluation of power, bias, and solution propriety. *Educational and Psychological Measurement, 73*(6), 913–934. https://doi.org/10.1177/0013164413495237

Wolff, H. G., & Preising, K. (2005). Exploring item and higher order factor structure with the Schmid-Leiman solution: Syntax codes for SPSS and SAS. *Behavior Research Methods, 37*(1), 48-58. https://doi.org/10.3758/BF03206397

Wood, J. M., Tataryn, D. J., & Gorsuch, R. L. (1996). Effects of under- and overextraction on principal axis factor analysis with varimax rotation. *Psychological Methods, 1*(4), 254–265. https://doi.org/10.1037/1082-989X.1.4.354

Woods, C. M. (2006). Careless responding to reverse-worded items: Implications for confirmatory factor analysis. *Journal of Psychopathology and Behavioral Assessment, 28*(3), 189–194. https://doi.org/10.1007/s10862-005-9004-7

Worthington, R. L., & Whittaker, T. A. (2006). Scale development research: A content analysis and recommendations for best practices. *Counseling Psychologist, 34*(6), 806–838. https://doi.org/10.1177/0011000006288127

Wothke, W. (1993). Nonpositive definite matrices in structural modeling. In K. A. Bollen & J. S. Long (Eds.), *Testing structural equation models* (pp. 256–293). Sage.

Xia, Y., & Yang, Y. (2019). RMSEA, CFI, and TLI in structural equation modeling with ordered categorical data: The story they tell depends on the estimation methods. *Behavior Research Methods, 51*(1), 409–428. https://doi.org/10.3758/s13428-018-1055-2

Xiao, C., Bruner, D. W., Dai, T., Guo, Y., & Hanlon, A. (2019). A comparison of missing-data imputation techniques in exploratory factor analysis. *Journal of Nursing Measurement, 27*(2), 313–334. https://doi.org/10.1891/1061-3749.27.2.313

Ximénez, C. (2009). Recovery of weak factor loadings in confirmatory factor analysis under conditions of model misspecification, *Behavior Research Methods, 41*(4), 1038–1052. https://doi.org/10.3758/BRM.41.4.1038

Yakovitz, S., & Szidarovszky, F. (1986). *An introduction to numerical computation.* Macmillan.

Yong, A. G., & Pearce, S. (2013). A beginner's guide to factor analysis: Focusing on exploratory factor analysis. *Tutorials in Quantitative Methods for Psychology, 9*(2), 79–94. https://doi.org/10.20982/tqmp.09.2.p079

Yuan, K.-H. (2005). Fit indices versus test statistics. *Multivariate Behavioral Research, 40*(1), 115–148. https://doi.org/10.1207/s15327906mbr4001_5

Zhang, G. (2014). Estimating standard errors in exploratory factor analysis. *Multivariate Behavioral Research, 49*(4), 339–353. https://doi.org/10.1080/00273171.2014.908271

Zhang, G., & Browne, M. W. (2006). Bootstrap fit testing, confidence intervals, and standard error estimation in the factor analysis of polychoric correlation matrices. *Behaviormetrika, 33*(1), 61–74. https://doi.org/10.2333/bhmk.33.61

Zhang, G., & Preacher, K. J. (2015). Factor rotation and standard errors in exploratory factor analysis. *Journal of Educational and Behavioral Statistics, 40*(6), 579–603. https://doi.org/10.3102/1076998615606098

Zinbarg, R. E., Revelle, W., Yovel, I., & Li, W. (2005). Cronbach's α, Revelle's β, and Mcdonald's ωh: Their relations with each other and two alternative conceptualizations of reliability. *Psychometrika, 70*(1), 123–133. https://doi.org/10.1007/s11336-003-0974-7

Zoski, K. W., & Jurs, S. (1996). An objective counterpart to the visual scree test for factor analysis: The standard error scree. *Educational and Psychological Measurement, 56*(3), 443–451. https://doi.org/10.1177/0013164496056003006

Zwick, W. R., & Velicer, W. F. (1986). A comparison of five rules for determining the number of components to retain. *Psychological Bulletin, 99*(3), 432–442. https://doi.org/10.1037/0033-2909.99.3.432

Zygmont, C., & Smith, M. R. (2014). Robust factor analysis in the presence of normality violations, missing data, and outliers: Empirical questions and possible solutions. *Tutorial in Quantitative Methods for Psychology, 10*(1), 40–55. https://doi.org/10.20982/tqmp.10.1.p040

Index